U.S.NRC
United States Nuclear Regulatory Commission

Protecting People and the Environment

NUREG/CR-7161
SAND2010-3380P

Synthesis of Distributions Representing Important Non-Site-Specific Parameters in Off-Site Consequence Analyses

Office of Nuclear Regulatory Research

AVAILABILITY OF REFERENCE MATERIALS
IN NRC PUBLICATIONS

NRC Reference Material

As of November 1999, you may electronically access NUREG-series publications and other NRC records at NRC's Public Electronic Reading Room at http://www.nrc.gov/reading-rm.html. Publicly released records include, to name a few, NUREG-series publications; *Federal Register* notices; applicant, licensee, and vendor documents and correspondence; NRC correspondence and internal memoranda; bulletins and information notices; inspection and investigative reports; licensee event reports; and Commission papers and their attachments.

NRC publications in the NUREG series, NRC regulations, and Title 10, "Energy," in the *Code of Federal Regulations* may also be purchased from one of these two sources.
1. The Superintendent of Documents
   U.S. Government Printing Office Mail Stop SSOP
   Washington, DC 20402–0001
   Internet: bookstore.gpo.gov
   Telephone: 202-512-1800
   Fax: 202-512-2250
2. The National Technical Information Service
   Springfield, VA 22161–0002
   www.ntis.gov
   1–800–553–6847 or, locally, 703–605–6000

A single copy of each NRC draft report for comment is available free, to the extent of supply, upon written request as follows:
Address: U.S. Nuclear Regulatory Commission
         Office of Administration
         Publications Branch
         Washington, DC 20555-0001
E-mail: DISTRIBUTION.RESOURCE@NRC.GOV
Facsimile: 301–415–2289

Some publications in the NUREG series that are posted at NRC's Web site address http://www.nrc.gov/reading-rm/doc-collections/nuregs are updated periodically and may differ from the last printed version. Although references to material found on a Web site bear the date the material was accessed, the material available on the date cited may subsequently be removed from the site.

Non-NRC Reference Material

Documents available from public and special technical libraries include all open literature items, such as books, journal articles, transactions, *Federal Register* notices, Federal and State legislation, and congressional reports. Such documents as theses, dissertations, foreign reports and translations, and non-NRC conference proceedings may be purchased from their sponsoring organization.

Copies of industry codes and standards used in a substantive manner in the NRC regulatory process are maintained at—
    The NRC Technical Library
    Two White Flint North
    11545 Rockville Pike
    Rockville, MD 20852–2738

These standards are available in the library for reference use by the public. Codes and standards are usually copyrighted and may be purchased from the originating organization or, if they are American National Standards, from—
    American National Standards Institute
    11 West 42nd Street
    New York, NY 10036–8002
    www.ansi.org
    212–642–4900

Legally binding regulatory requirements are stated only in laws; NRC regulations; licenses, including technical specifications; or orders, not in NUREG-series publications. The views expressed in contractor-prepared publications in this series are not necessarily those of the NRC.

The NUREG series comprises (1) technical and administrative reports and books prepared by the staff (NUREG–XXXX) or agency contractors (NUREG/CR–XXXX), (2) proceedings of conferences (NUREG/CP–XXXX), (3) reports resulting from international agreements (NUREG/IA–XXXX), (4) brochures (NUREG/BR–XXXX), and (5) compilations of legal decisions and orders of the Commission and Atomic and Safety Licensing Boards and of Directors' decisions under Section 2.206 of NRC's regulations (NUREG–0750).

United States Nuclear Regulatory Commission

Protecting People and the Environment

NUREG/CR-7161
SAND2010-3380P

Synthesis of Distributions Representing Important Non-Site-Specific Parameters in Off-Site Consequence Analyses

Manuscript Completed: October 2012
Date Published: April 2013

Prepared by:
N. E. Bixler, E. Clauss, and C. W. Morrow

Sandia National Laboratories
Albuquerque, New Mexico 87185-0734

J. A. Mitchell, C. Navarro, J. Barr, NRC Project Managers

NRC Job Codes N6159, Y6628, and V6227

Office of Nuclear Regulatory Research

ABSTRACT

The United States and the Commission of European Communities conducted a series of expert elicitations to obtain distributions for uncertain variables used in health consequence analyses related to accidental release of nuclear material. The distributions reflect degrees of belief for non-site-specific parameters that are uncertain and are likely to have significant or moderate influence on the results. The present work presents the effort to develop ranges of values and degrees of belief that fairly represent the divergent opinions of the experts while maintaining the resulting parameters within physical limits. Where necessary, there is a discussion of correlation coefficients that should be included when the uncertainty is used in a calculation. The methodology used a resampling of the experts' values and was based on the assumption of equal weights of the experts' opinions. Various statistical properties of the distributions, the median, the mean, and the mode, are presented so that the user can choose a parameter value when only a point estimate is desired.

TABLE OF CONTENTS

LIST OF FIGURES

LIST OF FIGURES (continued)

LIST OF FIGURES (continued)

LIST OF FIGURES (continued)

LIST OF FIGURES (continued)

LIST OF FIGURES (continued)

LIST OF FIGURES (continued)

LIST OF FIGURES (continued)

LIST OF FIGURES (continued)

LIST OF FIGURES (continued)

LIST OF TABLES

EXECUTIVE SUMMARY

The U.S. Nuclear Regulatory Commission (NRC) developed the MELCOR Accident Consequence Code System, Version 2, (MACCS2) specifically to evaluate offsite consequences from a hypothetical release of radioactive material into the atmosphere. The code models atmospheric transport and dispersion, emergency response actions, exposure pathways, health effects, and economic costs. There are a large number of input parameters required for the MACCS2 code, some of which have a range of possible values and varying level of uncertainty. This document presents distributions developed to provide analysts using MACCS2 guidance on the most appropriate values to use for these parameters.

The NRC and the Commission of European Communities (CEC) conducted a series of studies with panels of experts from the United States and the CEC (References 1 through 6). During those studies, the experts were asked to provide their own ranges of values and degrees of belief for a large number of input parameters used in offsite consequence codes. The parameters were those that the authors of the studies decided were (1) uncertain in value, (2) likely to be significant or moderate in their influence on the calculated results, and (3) applicable to all sites (that is, not site-specific).

The present study, performed by Sandia National Laboratories, uses the results of the previous studies by combining the varying expert opinions of each parameter to form a single distribution for each parameter. The distribution for each parameter is developed in a way that fairly represents the experts' divergent opinions while maintaining the parameter within physical limits. A resampling method was used with equal weight assigned to each expert. These distributions provide guidance to analysts using the MACCS2 code regarding the input values that should be used for these parameters.

In several cases, correlation coefficients are necessary to maintain physical relationships among groups of parameters when the full distribution is used in an uncertainty study. The suggested correlation coefficients, where necessary, are discussed at the end of each section.

Different offsite consequence measures (e.g., early fatality risk, latent cancer fatality risk, or costs) are dependent in different ways on the input parameters evaluated. For use in consequence analyses in which only a point estimate is desired, the medians, means, and modes of the distributions are provided to allow the user to select an appropriate value. If results at one or more specific quantile levels are desired, it is preferable to conduct an uncertainty study in which all of the relevant parameters are sampled over their entire distributions and the results are evaluated at each of the appropriate quantile levels. MACCS2, and its graphical user interface and preprocessor, WinMACCS, have been tailored to perform such evaluations.

ACKNOWLEDGMENTS

The authors thank Jocelyn Mitchell for initiating and supporting this work, for offering guidance as the work progressed, and for providing valuable comments on the contents of this report. We deeply regret that she was not able to see the completion of this work.

ACRONYMS AND INITIALISMS

CDF	Cumulative Distribution Function
CEC	Commission of European Communities
COSYMA	COde SYstem from MAria
D_{50}	Dose at which 50% of the public experience a health effect
EC	European Commission
LD_{50}	Lethal Dose for 50% of the public
LHS	Latin Hypercube Sampling
MACCS2	MELCOR Accident Consequence Code System, Version 2
NRC	U.S. Nuclear Regulatory Commission
PDF	Probability Distribution Function
RSM	ReSampling Method
SOARCA	State-of-the-Art Reactor Consequence Analyses project

1.0 BACKGROUND

An expert elicitation was conducted jointly by the United States Nuclear Regulatory Commission (NRC) and the Commission of the European Communities (CEC) to obtain distributions for uncertain variables used in health consequence analyses related to accidental release of nuclear materials into the atmosphere. The distributions reflect degrees of belief for parameters that are uncertain and are likely to be significant or moderate in their influence on the calculated results. The results of the elicitation are given in References 1 through 6. The references explain that the approach was jointly formulated and was based on two important ground rules: (1) the current code models would not be changed because both the NRC and European Commission (EC) were interested in the uncertainties in the predictions produced by MACCS (the NRC's off-site consequence code) [Reference 7] and the Code System from Maria (COSYMA) (the EC's off-site consequence code), respectively, and (2) the experts would be asked only to assess physical quantities that hypothetically could be measured in experiments. The reasons for these ground rules are that (1) the codes have already been developed and applied in the US and the EC risk assessments, and (2) eliciting physical quantities avoids ambiguity in definitions of variables; more important, the physical quantities elicited are not tied to any particular model, and thus have a much wider potential application. The study involved several phases: preparation stage, expert training meetings, preparation of the assessments and written rationale, expert elicitation sessions, and processing of the elicited results. The values elicited from each of the experts for each of the parameters are given in the references.

This present work represents the effort to develop ranges of values and degrees of belief that fairly represent the divergent opinions of the experts, while maintaining the resulting parameters within physical limits. The development is performed specifically with the MACCS2 code in mind; only the set of parameters that can be used as MACCS2 input parameters is evaluated in this report.

The purpose for this work is to facilitate a more comprehensive evaluation of uncertainty in future consequence analyses. Typically, evaluation of uncertainty in consequence analysis has been limited to the influence of weather, a type of aleatory uncertainty, but has not been extended to other input parameters that are not well quantified. The purpose of this report is to assimilate distributions for important parameters that are site independent and are not well quantified so that future consequence analyses can more thoroughly assess the inherent uncertainty, both from weather and from other sources of uncertainty related to lack of knowledge. Specifically, the distributions developed in this report are intended to be used in the State-of-the-Art Reactor Consequence Analysis (SOARCA) uncertainty analysis, which is being performed by Sandia National Laboratories for the NRC.

Within each of References 1 through 6, each expert who participated in the process is assigned a letter in order to preserve confidentiality. In most cases, a subset of the experts provided an opinion on a specific parameter. The relationship between the expert and the letter used to designate that expert is preserved throughout the original documents. The same letter designations are used in this report as well. Thus, the set of letters representing the experts' opinions in Sections 2 through 7 of this report differ from plot to plot, depending on which subset of experts provided an opinion on the parameter shown in the plot.

Some previous work has been done along the lines of the work presented here. Wheeler et al. [Reference 8] evaluated the uncertainty of a set of parameters relevant to the safety analysis of the Cassini Mission. This effort did not consider all of the parameters needed for consequence codes like MACCS2 and only considered a few percentile levels (5%, 50%, and 95%). Furthermore, this study did not attempt to provide correlations of input parameters, such as aerosol size, to the uncertain outputs, such as deposition velocity. In these regards, the current study is more comprehensive than the one conducted by Wheeler et al. and should prove to have a broader range of applicability.

1.1   Methodology

Some expert data were elicited for the following quantile values: 0.00, 0.05, 0.50, 0.95, and 1.00; however, most expert data were only elicited for the 0.05, 0.50, and 0.95 quantiles. In this work, we take each elicited opinion as being equally likely since we have no basis for considering the opinion of one expert more credible than another. This means that each expert's data are represented by an equal number of resampled points. While equal weighting seems straightforward on the surface, it is somewhat more complicated than might be expected in cases where some experts provided values for the 0.00 and 1.00 quantiles while other experts did not.

The initial concept for treating nonhomogeneous expert data was to normalize over quantile ranges so that each quantile range would receive the same number of data points. This concept results in more resampling points per expert for the extreme ranges, from 0.00 to 0.05 and 0.95 to 1.00, than for the intermediate range, 0.05 to 0.95, since only a small number of experts contributed values in the extreme ranges and most contributed values in the intermediate range. To illustrate this point, consider a case of eight experts providing values for the 0.05, 0.5, and 0.95 quantiles, but a subset of three providing values for the 0 and 1 quantile levels. If eight resample points were evaluated for each of the three experts providing opinions in the ranges 0.00 to 0.05 and 0.95 to 1.00 and three resample points were chosen for each expert from 0.05 to 0.95, then the result is that all quantile ranges are represented by the same density of resample points and all experts are treated equally over ranges where they supplied data. However, the values of the three experts have a greater weight in the extreme ranges.

While the initial concept seems like a very good one, it has one fatal shortcoming for the analysis at hand. Only in the report on dispersion and deposition were the experts asked to provide 0.00 and 1.00 quantile values; in the other reports, the experts were not asked to provide these values. In these cases, the initial resampling concept would only be able to produce distributions over the 0.05 to 0.95 quantile range, which is unacceptable because sampling algorithms (e.g., standard Monte Carlo or Latin Hypercube Sampling (LHS)) require a complete probability distribution to be defined.

The second concept for treating nonhomogeneous expert data is to estimate the missing values using some simple algorithm. The algorithm that was adopted was to divide the expert's 0.05 quantile value by 2 to get the 0.00 quantile value and to multiply the expert's 0.95 quantile value by 2 to get the 1.00 quantile value. The factor of two was chosen because it approximately fit the values for the experts who did provide the 0.00 and 1.00 quantile values. The second concept seems a bit more arbitrary than the first concept; it was chosen here because it could be extended to the evaluation of all the expert elicitation data.

A second choice that had to be made in order to perform the resampling is the type of interpolation to be used to represent the expert data. The choice here is the simplest one, linear interpolation. Once the resampled data are created, they are ordered and assigned a quantile value. This process is illustrated in Tables 1-1 and 1-2 for a simplified example in which there are only two experts. For some of the parameters, data at a given quantile level are fitted using linear regression to calculate the coefficients used to describe the overall trends. The results are provided in the following sections.

In each of the subsequent sections, results are tabulated at a set of quantile levels: 0.00, 0.01, 0.05, 0.10, 0.25, 0.50, 0.75, 0.90, 0.95, 0.99, and 1.00. In addition, means and modes are included in the tabulations. Means are computed simply as the arithmetic average of the set of resampled values. Calculation of the modes is described below.

The mode corresponds to the peak in the probability distribution function (PDF). The PDF is the derivative of the cumulative distribution function (CDF). Problems arise in the calculation of the PDF and the determination of the mode when the CDF is not smooth. Some compromises must be made to determine the PDF and its peak, the mode, in a reasonable fashion. This is illustrated in the remainder of this section.

Table 1-2 shows an example of the method used to determine the mode of a distribution. The third column shows the values used to determine the mode (i.e., the derivative of the CDF) which is calculated using the following formula with $m = 8$:

$$M_{n,m} = \frac{\sum\limits_{i=1,m} (Q_{n+i} - Q_{n-i})}{\sum\limits_{i=1,m} (D_{n+i} - D_{n-i})} \tag{1.1}$$

where
n	=	a subscript representing the point being evaluated
m	=	an integer representing the number of points to be included in the difference
i	=	a dummy index
$M_{n,m}$	=	is a finite difference approximation to a derivative
Q	=	a quantile value
D	=	a resampled data point from the original expert data

The mode of the distribution shown in Table 1-2 corresponds to the peak value of M, which is 4.691.

Table 1-1: Resampled data for two experts.

Quantile	Expert	
	A	B
0.0	4.300	3.784
0.1	4.477	4.520
0.2	4.522	4.606
0.3	4.566	4.691
0.4	4.611	4.776
0.5	4.657	4.862
0.6	4.729	4.953
0.7	4.792	5.033
0.8	4.865	5.124
0.9	4.937	5.215
1.0	5.667	5.954

Table 1-2: Example of merged and sorted data and estimation of mode.

Quantile (Q)	Ordered Data (D)	M
0.000	3.784	
0.048	4.300	
0.095	4.477	
0.143	4.520	
0.190	4.522	
0.238	4.566	
0.286	4.606	
0.333	4.611	
0.381	4.657	1.065
0.429	4.691	1.276
0.476	4.729	1.274
0.524	4.776	1.191
0.571	4.792	.0980
0.619	4.862	0.794
0.667	4.865	
0.714	4.937	
0.762	4.953	
0.810	5.033	
0.857	5.124	
0.905	5.215	
0.952	5.667	
1.000	5.954	

4

The summation in Equation (1.1) can be over a single point or over a large number of points, depending on the selection of the value of m. Few points provide large fluctuations in the value of M due to noise in the data. A large number smooths the result but looses the local character of the data. After evaluating values of m from 2 to 10, $m = 8$ was selected for this work.

The difficulties associated with finding the mode of a resampled distribution are illustrated in Figure 1.1 for crosswind dispersion under stability classes A and B. Because of the interpolation scheme used to represent the few data points provided by each of the experts, the data used for resampling have discontinuities in slope. When all of the experts' data are assembled, the slope has a number of peaks and valleys, as illustrated in the figure.

When $m = 2$, the calculation of the slope is extremely noisy, as shown in the first plot in Figure 1.1. The bright pink square is the location of the tallest peak in the plot and would correspond to the mode if this method were used to estimate the PDF. The second plot uses $m = 4$ and is clearly much smoother than the one for $m = 2$, but still fairly noisy. Notice that the location of the mode in this plot is dramatically different than in the plot for $m = 2$. The plot with $m = 6$ is smoother yet, but produces an estimate for the mode that is identical to the previous plot. This trend continues for $m = 8$ and $m = 10$. The estimated values of the mode are nearly the same, but not identical for $m = 4$ through 10.

A choice of either $m = 8$ or 10 would have provided reasonable estimates of the mode in the subsequent sections. The choice of $m = 8$ was selected after evaluating the experts' data for a number of parameters. This choice was used consistently throughout each of the subsequent sections.

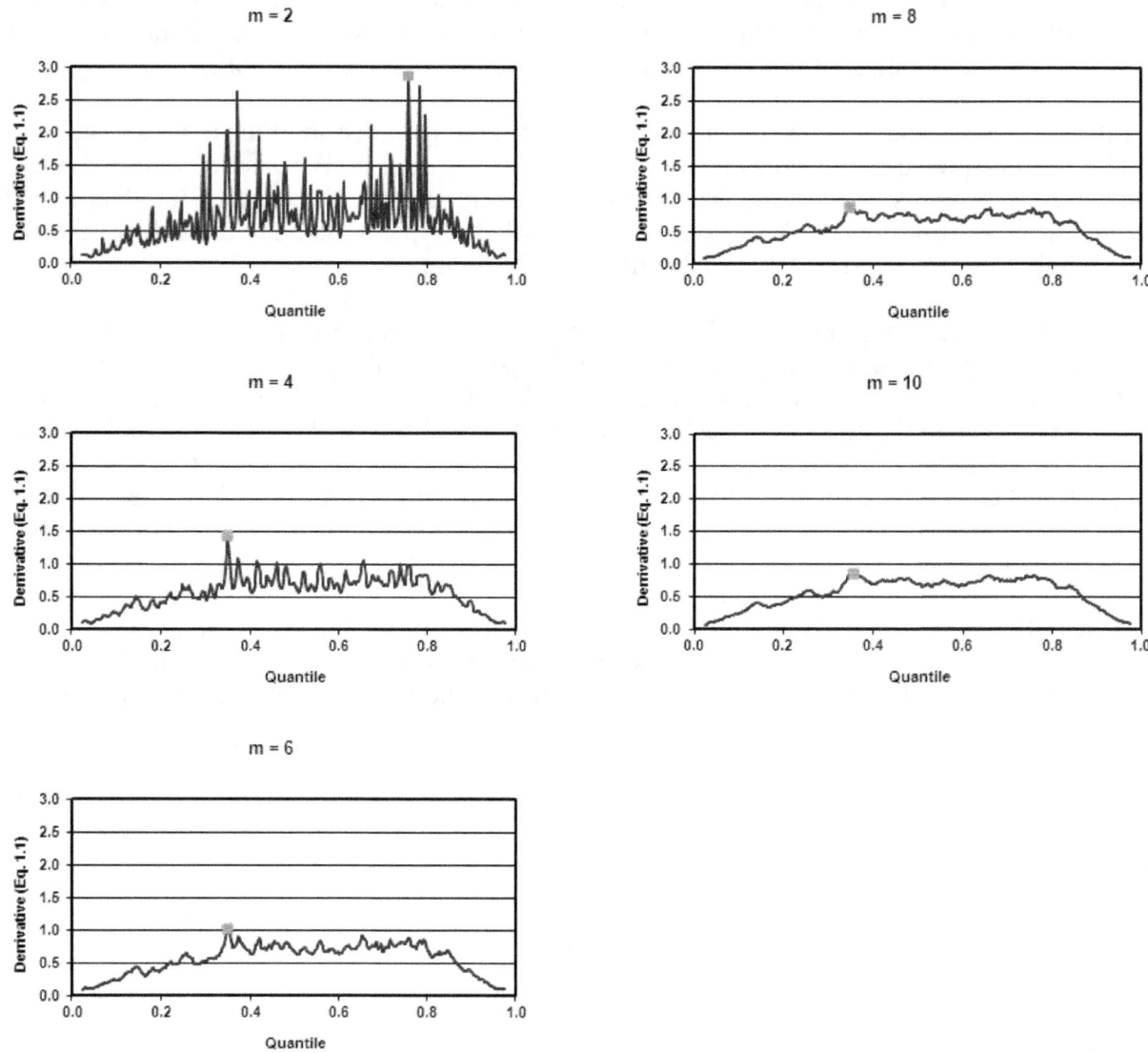

Figure 1.1.   Illustration of the evaluation of a PDF and mode (shown as a bright pink square) of the distribution using *m* = 2, 4, 6, 8, and 10 in Equation (1.1)

2.0 EVALUATION OF DISPERSION DATA

MACCS2 allows two approaches for evaluating dispersion as a function of downwind distance. The first, and probably the more widely used, is based on a lookup table provided by the user. However, this approach is inconvenient for representing uncertain dispersion values. The second approach is much more convenient for expressing degree of belief and uses power-law equations for σ_y and σ_z, which are defined as follows:

$$\sigma_y = a\left(\frac{x}{x_0}\right)^b \tag{2.1}$$

$$\sigma_z = c\left(\frac{x}{x_0}\right)^d \tag{2.2}$$

where
σ_y	=	crosswind dispersion parameter (m)
σ_z	=	vertical dispersion parameter (m)
a	=	linear coefficient for crosswind dispersion (m)
c	=	linear coefficient for vertical dispersion (m)
x	=	downwind distance (m)
x_0	=	downwind distance scale, 1 m
b	=	exponential parameter for crosswind dispersion (dimensionless)
d	=	exponential parameter for vertical dispersion (dimensionless)

Here, distributions for a, b, c, and d are calculated to reflect the degree of belief expressed during the expert elicitation process. The distributions are calculated for Pasquill-Gifford stability classes A/B, C, D, and E/F. The basic methodology used to calculate the coefficients in Equations (2.1) and (2.2) begins by resampling the expert data for each expert.

A final difficulty with the expert data arose in evaluating the parameters associated with vertical dispersion, c and d, because the experts were not asked to provide σ_z directly. Instead, the experts were asked to provide the ratio of off-centerline to centerline concentrations. Standard equations for a Gaussian plume, which assume reflective planes at the ground surface and at the mixing layer height, are used to calculate σ_z given concentration ratio for values between 0 and 1, not including the endpoints. The value of σ_z approaches 0 when the concentration ratio approaches 0 and it approaches infinity when the concentration ratio approaches 1. Values of this ratio greater than 1 do not have any physical meaning in the context of a Gaussian plume model.

The Gaussian plume equation is [Reference 9]:

$$\frac{\Psi}{Q} = \frac{1}{2\pi\sigma_y\sigma_z u}\exp\left[-\frac{1}{2}\left(\frac{y}{\sigma_y}\right)^2\right] \sum_{n=-\infty}^{\infty} \exp\left[-\frac{1}{2}\left(\frac{2nh-H-z}{\sigma_z}\right)^2\right] + \exp\left[-\frac{1}{2}\left(\frac{2nh+H-z}{\sigma_z}\right)^2\right] \tag{2.3}$$

7

where

Ψ =   concentration of released contaminant (kg/m^3 or Bq/m^3)
Q =   source rate of released contaminant (kg/s or Bq/s)
u =   wind speed at plume centerline (m/s)
n =   a dummy index used for summation
y =   crosswind distance from the plume centerline (m)
z =   vertical distance from the ground (m)
h =   height of the mixing layer (m), which is taken to be 1500 m
H =   height of the plume centerline (m), which is given as 10 m

From Equation (2.3), the ratio of the concentration at a vertical elevation z to the centerline concentration is

$$\frac{\Psi_z}{\Psi_{CL}} = \frac{\sum\limits_{-\infty}^{\infty}\exp\left[-\frac{1}{2}\left(\frac{2nh-H-z}{\sigma_z}\right)^2\right]+\exp\left[-\frac{1}{2}\left(\frac{2nh+H-z}{\sigma_z}\right)^2\right]}{\sum\limits_{-\infty}^{\infty}\exp\left[-\frac{1}{2}\left(\frac{2nh-2H}{\sigma_z}\right)^2\right]+\exp\left[-\frac{1}{2}\left(\frac{2nh}{\sigma_z}\right)^2\right]} \tag{2.4}$$

Equation (2.4) is used to calculate values of σ_z given expert values of Ψ_z/Ψ_{CL}. This calculation is done iteratively because of the highly nonlinear dependence on σ_z.

Figure 2-1 shows the dependence of Ψ_z/Ψ_{CL} on σ_z when $z = 60$ m and $H = 0$ m. This figure clearly depicts the difficulty of calculating σ_z given a value of Ψ_z/Ψ_{CL} close to unity. In fact, there is no unique value of σ_z when Ψ_z/Ψ_{CL} is unity. The statement that Ψ_z/Ψ_{CL} is approximately unity (i.e., rounds to unity for some finite number of significant digits) implies that σ_z must be greater than or equal to some minimum value. For example, using the parameters chosen to generate Figure 2-1, $\sigma_z > 180$ m if Ψ_z/Ψ_{CL} is 1.0, $\sigma_z > 580$ m if Ψ_z/Ψ_{CL} is 1.00, and $\sigma_z > 1240$ m if Ψ_z/Ψ_{CL} is 1.000.

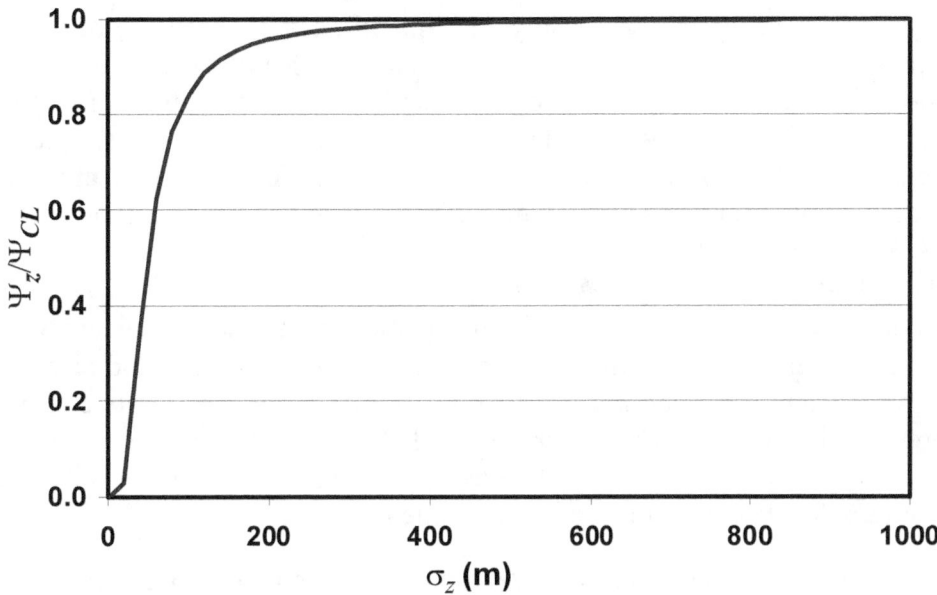

Figure 2-1. **Illustration of the functional dependence of Ψ_z/Ψ_{CL} on σ_z given in Equation (2.4)**

Unfortunately, a significant number of the elicited values for σ_z were 1. Depending on the expert, the elicited values were provided with 1, 2, or 3 significant digits. The most likely explanation for concentration ratios of 1 is simply that, to the number of significant digits provided by the experts, the number rounded to 1. As explained above, this corresponds only to a lower bound on the value of σ_z. Even worse, a number of experts provided concentration ratios greater than 1.

The approach adopted was to exclude data for σ_z if the expert provided concentration ratios greater than 1 because such values are inadmissible with a Gaussian plume model. The choice of which experts to include was done by stability class (i.e., an expert was excluded in the evaluation of σ_z for a stability class if any of his values exceeded 1). Experts who provided values of 1 for the concentration ratio were not excluded. Instead, a simple algorithm was adopted to estimate σ_z in these cases. If the concentration ratio of 1 corresponded to the 0.00 or 1.00 quantile, then the factor-of-2 approach described above was used to fill in the value of σ_z. In other words, the value assigned to σ_z at the 0.00 quantile was half the value of σ_z at the 0.05 quantile; the value assigned to σ_z at the 1.00 quantile was twice the value of σ_z at the 0.95 quantile. In several cases, experts provided concentration ratios of 1 for the 0.95 quantile. In one case, an expert provided values of 1 at all quantile levels. In general, values of 1 at the 0.95 quantile level were filled in by scaling the 0.50 quantile value by 1.5, a value that seemed to fit the overall trend. The one case in which an expert provided values of 1 at all quantile levels is discussed below in the section on the A/B stability class.

Initially, an attempt was made to develop distributions for all four of the coefficients (a, b, c, and d) in Equations (2.1) and (2.2). This turned out to generate distributions for b and d that were neither monotonically increasing nor decreasing with quantile level. This is permissible in principle because the parameter combinations for (a,b) and (c,d) can lead to a monotonic distribution of dispersion coefficients even if the correlations of b and d are not monotonic. However, since values of b and d must be sampled, the lack of montonicity is problematic. The simplest solution was to force c and d to have a single value for each stability class. This treatment, although less general, produced correlations that agreed very well with the original data, as demonstrated below.

2.1   Distributions for Stability Class A/B

Table 2-1 provides values for a, b, c, and d from Equations (2.1) and (2.2) when the Pasquill-Gifford stability class is either A or B (here, the two classes are combined and referred to as A/B). These values were calculated using linear regression to match the expert data as closely as possilbe. The figures in this section that show curves labeled "Fit" were generated using the parameter values in this table. To interpolate values between the quantile levels provided in the table, it is preferable to interpolate the logarithms of a and c linearly.

Table 2-1:   Values for cross-wind and vertical dispersion coefficients for Pasquill-Gifford stability class A/B.

Quantile	Crosswind Dispersion Parameters		Vertical Dispersion Parameters	
	a (m)	b (dimensionless)	c (m)	d (dimensionless)
0.00	0.0650	0.866	0.0056	1.277
0.01	0.1515	0.866	0.0089	1.277
0.05	0.2586	0.866	0.0132	1.277
0.10	0.3381	0.866	0.0166	1.277
0.25	0.4861	0.866	0.0252	1.277
0.50	0.7507	0.866	0.0361	1.277
0.75	1.1379	0.866	0.0598	1.277
0.90	1.6222	0.866	0.0800	1.277
0.95	2.0731	0.866	0.0961	1.277
0.99	3.2179	0.866	0.1336	1.277
1.00	4.0698	0.866	0.1951	1.277
Mean	0.7393	0.866	0.0370	1.277
Mode	0.6282	0.866	0.0401	1.277

2.1.1 Crosswind Dispersion

The crosswind dispersion values specified in the expert elicitation are in terms of σ_y. The MACCS2 code, which uses a polar coordinate system, requires crosswind dispersion values in terms of σ_θ, which is expressed in units of distance along an arc at a specified radius rather than along lines perpendicular to the wind direction. The physical picture is illustrated in Figure 2-2. In the past, no distinction was made between σ_y and σ_θ when defining the input parameters for MACCS2 (i.e., σ_θ was simply set equal to σ_y). Not making this distinction is acceptable provided that crosswind dispersion, σ_y, is significantly less than the downwind distance, x; this distinction is important when crosswind dispersion is similar to downwind distance; it is crucial when crosswind dispersion is significantly greater than downwind distance. Values of σ_y are converted to values of σ_θ as follows:

$$\sigma_\theta = x \cdot \mathrm{atan}(\sigma_y / x) \qquad (2.5)$$

The angle in a polar coordinate system is $\mathrm{atan}(\sigma_y / x)$, which from Equation (2.5) equals σ_θ / x. Equation (2.5) defines σ_θ so that the point $(x, \sigma_\theta / x)$ in a polar coordinate system lies on a straight line from the origin to the point (x, σ_θ) in a cartesian coordinate system. Notice that the polar formulation is not identical to the original cartesian formulation. In fact, there is no simple way to maintain consistency between the two coordinate systems. The scheme adopted is certainly preferable to simply setting the value of σ_θ equal to σ_y. To see that this is so, imagine the case where σ_y is much greater than the downwind distance, x. In a cartesian coordinate system, the point (x, σ_θ) always lies downwind of the point of release (i.e., to the right of the y axis). In the polar coordinate system, the point $(x, \sigma_\theta / x)$ would approach the y axis if σ_θ is defined by Equation (2.5); on the other hand, the point $(x, \sigma_y / x)$ in a polar coordinate system could lie anywhere on a circle of radius x, including points which are upwind of the point of release. This latter picture makes no physical sense and should be rejected.

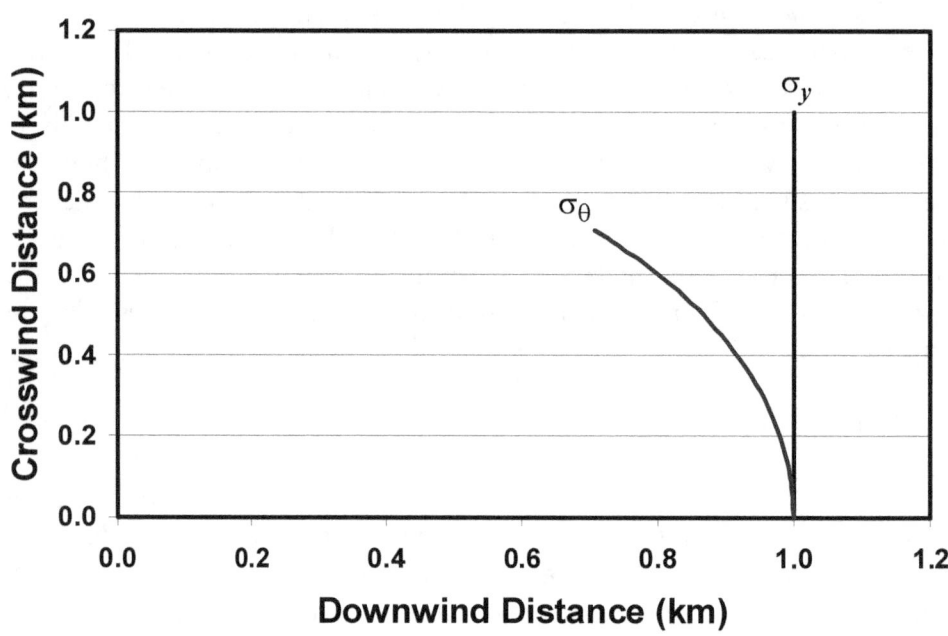

Figure 2-2. **Illustration of differences between crosswind dispersion in cartesian and polar coordinate systems**

Figures 2-3 through 2-6 display the expert data for each of the eight experts and the resampled data (labeled RSM throughout the report for resampling method) that were created for this study. Missing values in the expert data are determined using the algorithm described in Section 1. The original data for σ_y are converted to σ_θ using Equation (2.5). The figures show the resampled data span the entire ranges of values provided by the experts. This is a significant advantage over other methods that were considered, which all employed least-square fitting of the original data.

Figures 2-7 and 2-8 compare the final results tabulated in Table 2-1 with the resampled data shown in Figures 2-3 through 2-6. In Figure 2-7, the comparisons are made at fixed quantile level. In Figure 2-8, the comparisons are made at fixed downwind distances. The agreement is generally quite good.

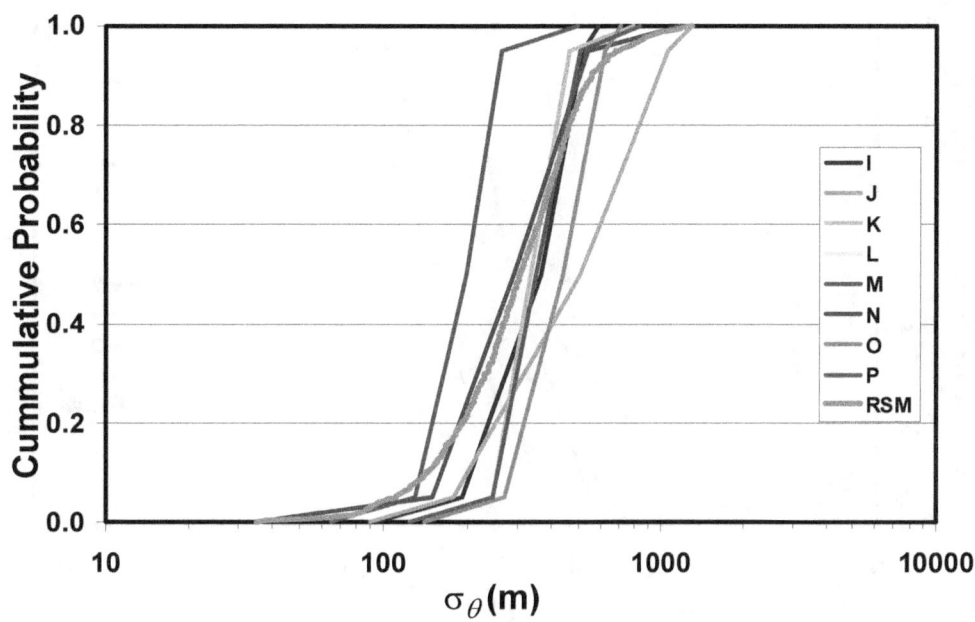

Figure 2-3. Expert and resampled data for σ_θ at 1 km downwind and Pasquill-Gifford stability class A/B

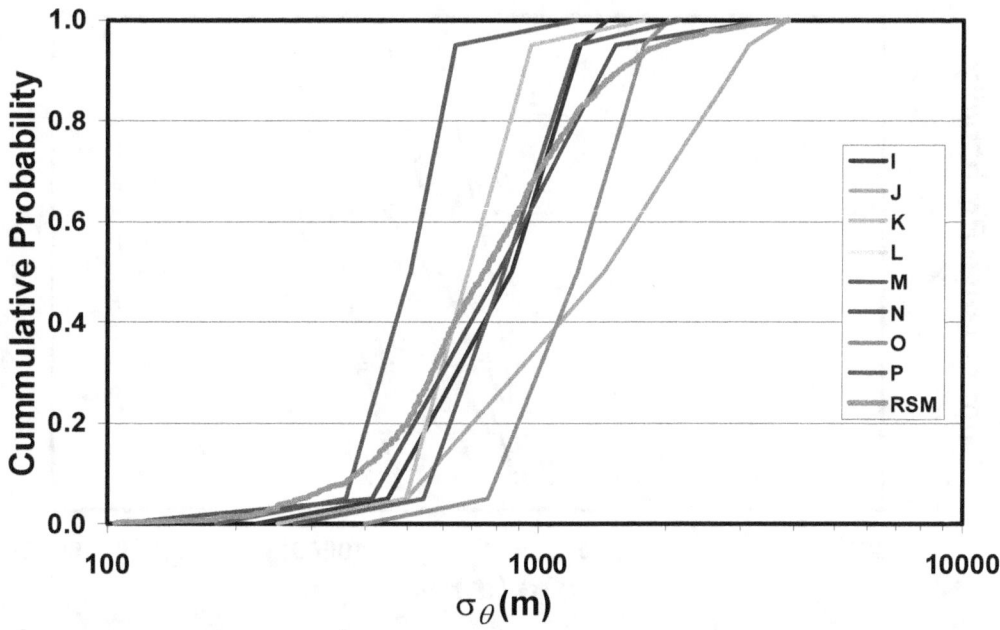

Figure 2-4. Expert and resampled data for σ_θ at 3 km downwind and Pasquill-Gifford stability class A/B

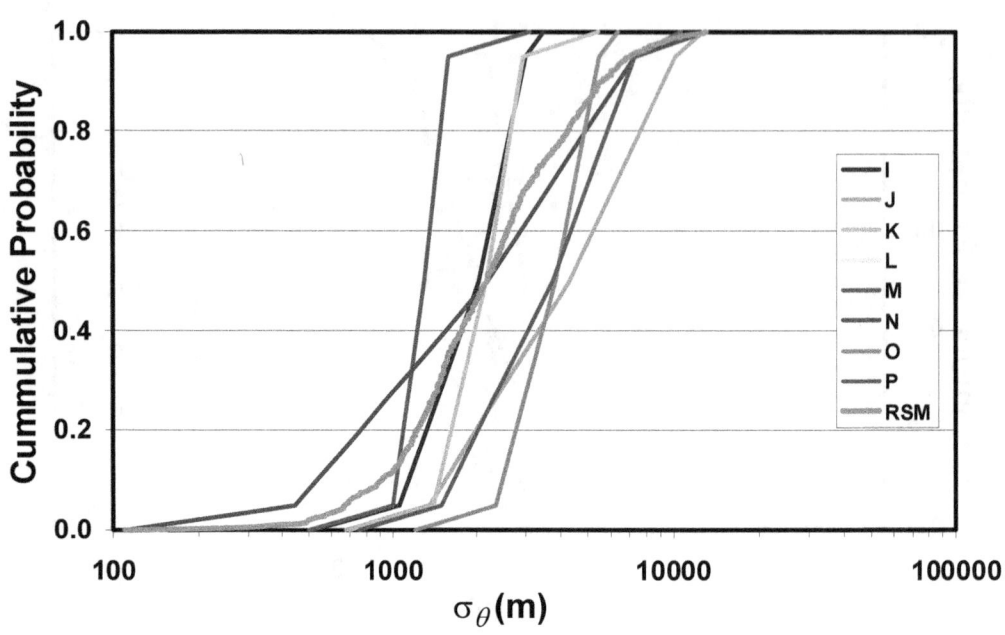

Figure 2-5. Expert and resampled data for σ_θ at 10 km downwind and Pasquill-Gifford stability class A/B

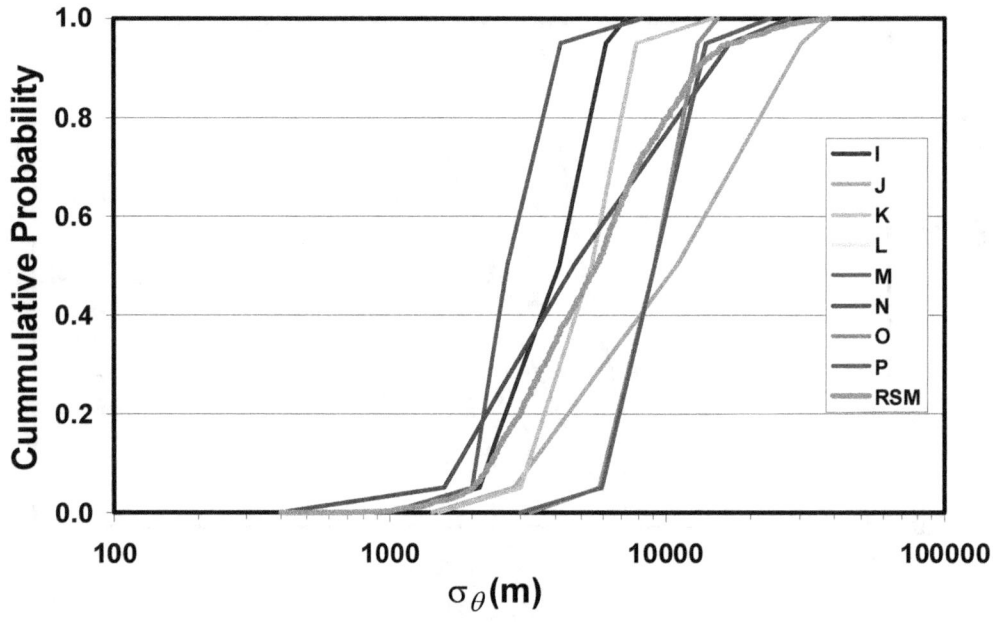

Figure 2-6. Expert and resampled data for σ_θ at 30 km downwind and Pasquill-Gifford stability class A/B

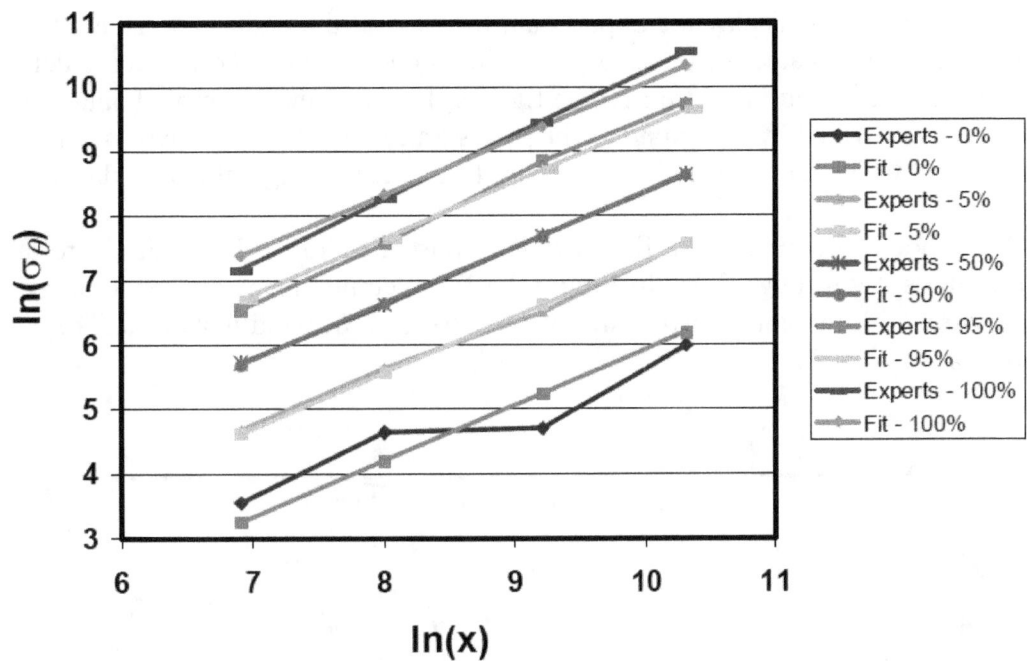

Figure 2-7. Comparison of resampled expert data for σ_θ and Pasquill-Gifford stability class A/B with the calculated values using the parameters in Table 2-1

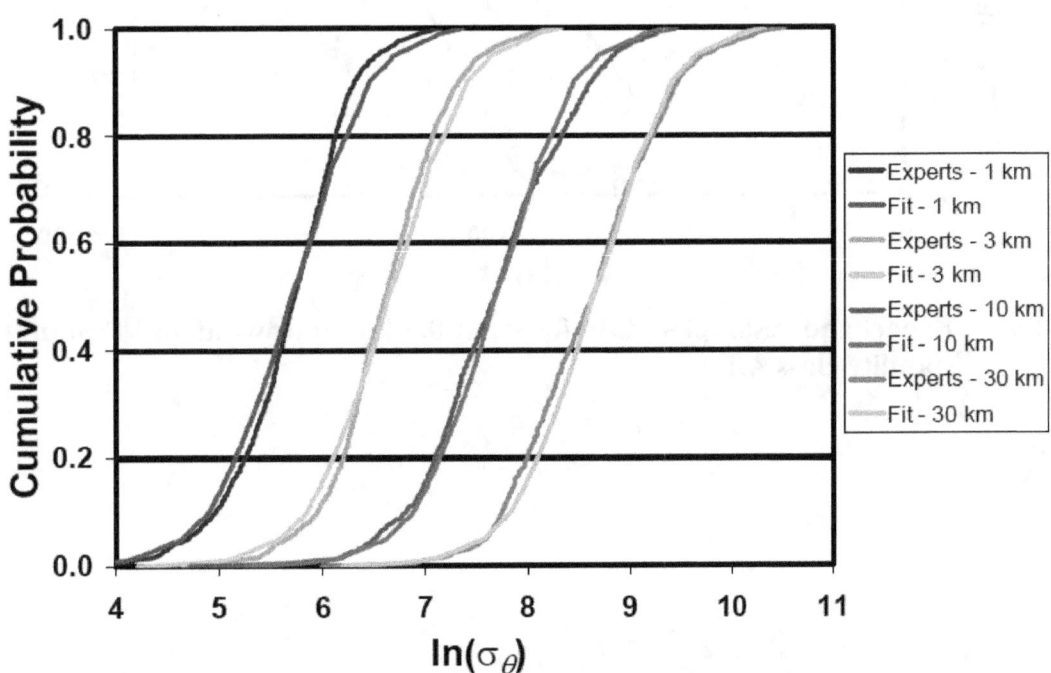

Figure 2-8. Comparison of resampled expert data for σ_θ and Pasquill-Gifford stability class A/B with the calculated values using the parameters in Table 2-1

2.1.2 Vertical Dispersion

Figures 2-9 through 2-11 display the expert data for each of the eight experts and the resampled data (RSM) that were created for this study. Missing values in the expert data are determined using the algorithm described in Section 1. The figures show that the resampled data span the entire ranges of values provided by the experts. This is a significant advantage over other methods that were considered, which all employed least-square fitting of the original expert data.

Figures 2-12 and 2-13 compare the final results reported in Table 2-1 with the resampled data shown in Figures 2-9 through 2-11. In Figure 2-12, the comparisons are made at fixed quantile level. In Figure 2-13, the comparisons are made at fixed downwind distances. The agreement is remarkably good.

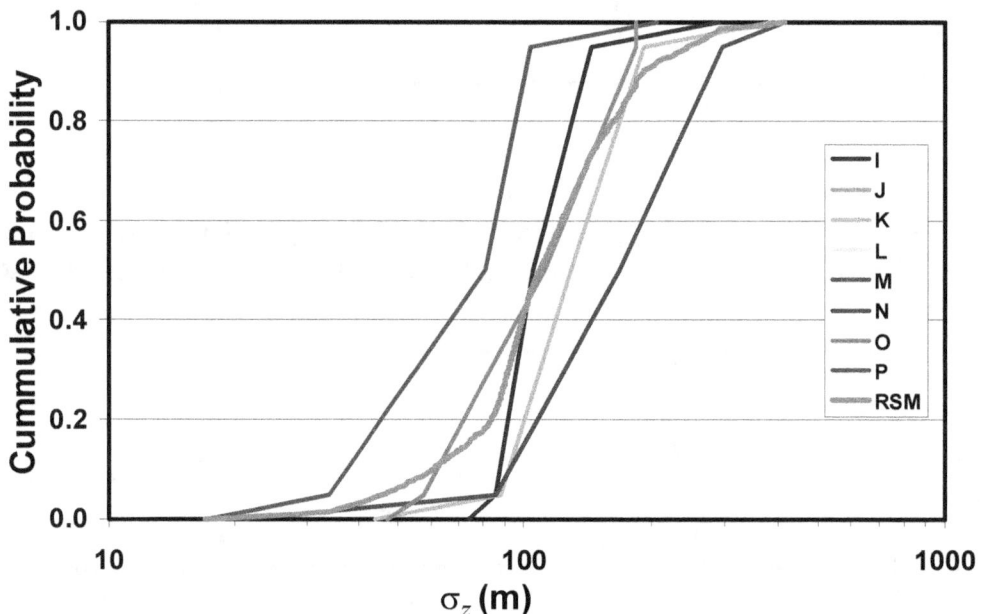

Figure 2-9. **Expert and resampled data for σ_z at 0.5 km downwind and Pasquill-Gifford stability class A/B**

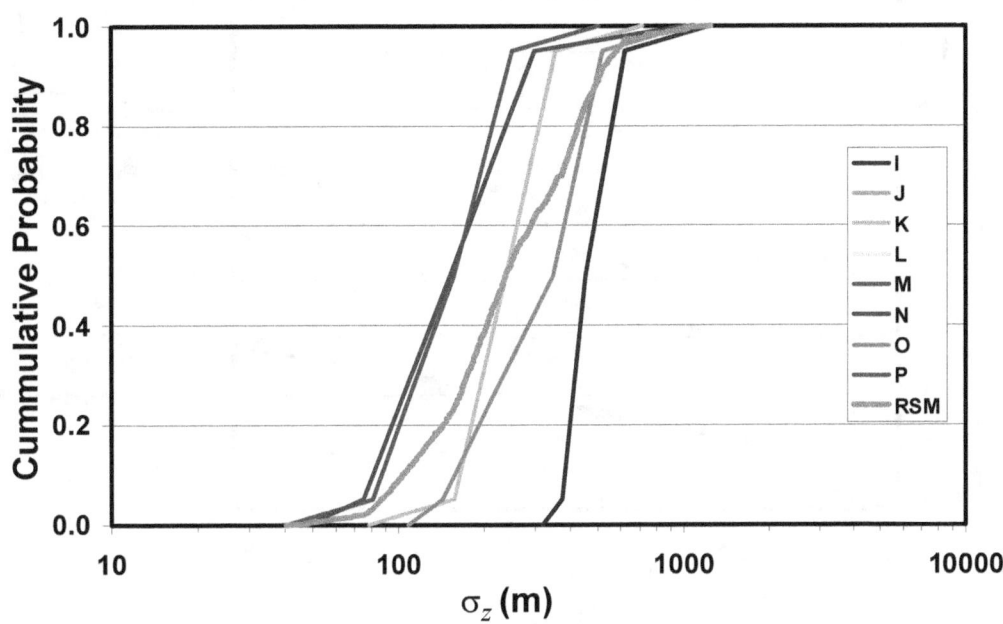

Figure 2-10. Expert and resampled data for σ_z at 1 km downwind and Pasquill-Gifford stability class A/B

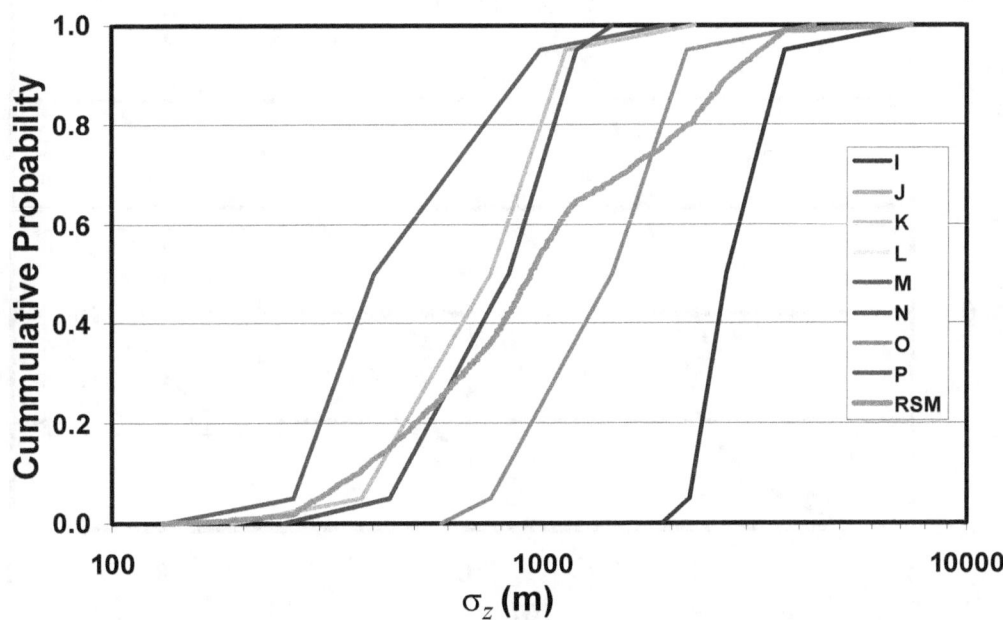

Figure 2-11. Expert and resampled data for σ_z at 3 km downwind and Pasquill-Gifford stability class A/B

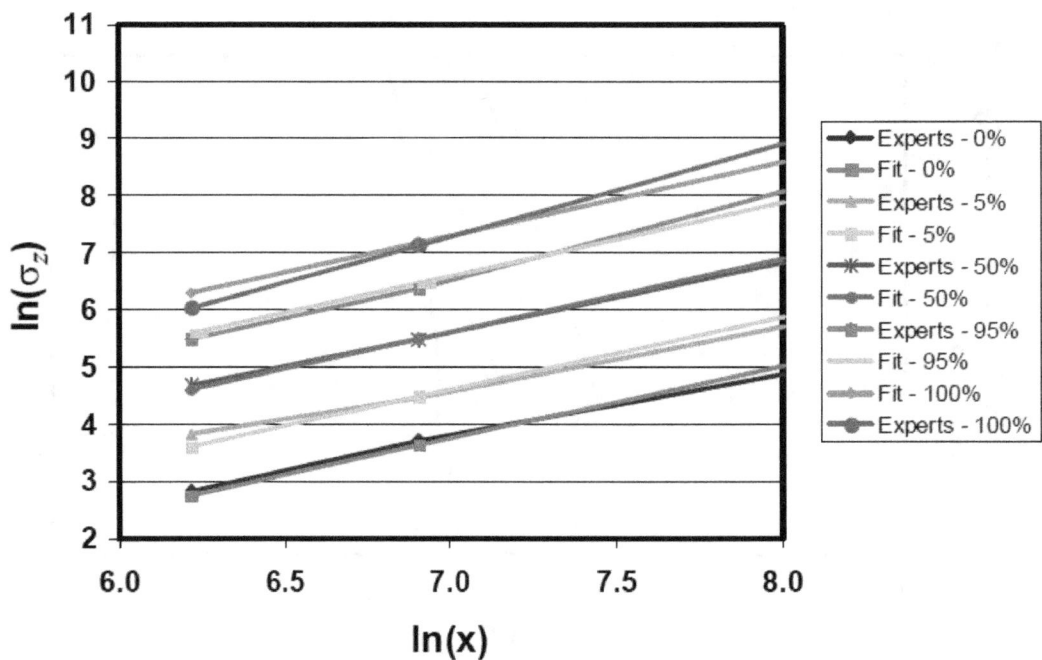

Figure 2-12. Comparison of resampled expert data for $\sigma_{\underline{z}}$ and Pasquill-Gifford stability class A/B with the calculated values using the parameters in Table 2-1

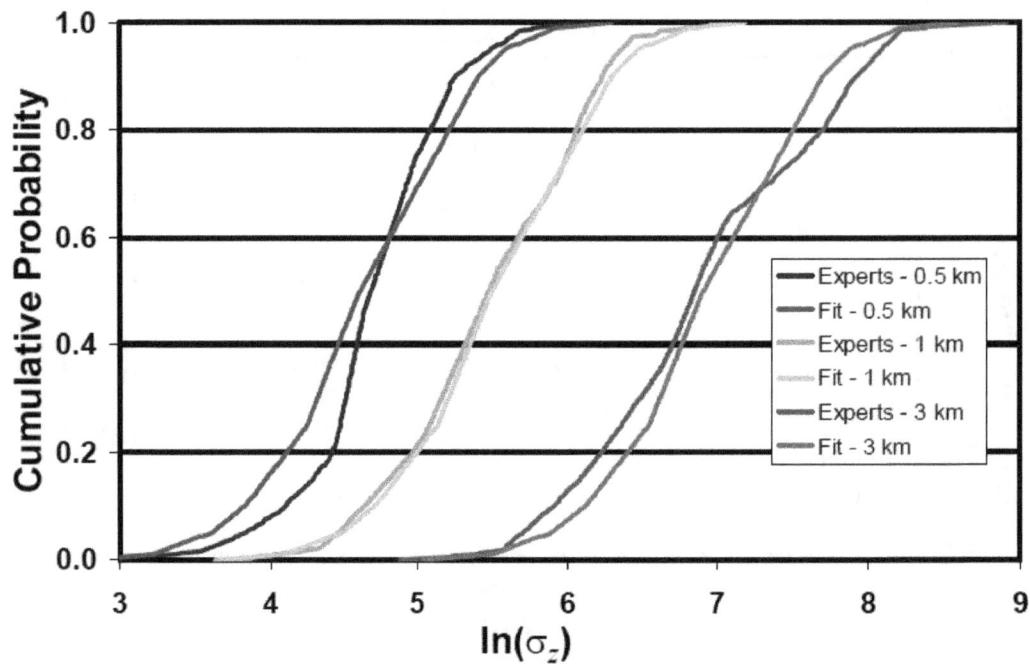

Figure 2-13. Comparison of resampled expert data for $\sigma_{\underline{z}}$ and Pasquill-Gifford stability class A/B with the calculated values using the parameters in Table 2-1

2.2 Distributions for Stability Class C

Table 2-2 provides values for a, b, c, and d from Equations (2.1) and (2.2) when the Pasquill-Gifford stability class is C. To interpolate values between the quantile levels provided in the table, it is preferable to interpolate the logarithms of a and c linearly.

2.2.1 Crosswind Dispersion

Figures 2-14 through 2-17 display the expert data for each of the eight experts and the resampled data that were created for this study. Missing values in the expert data are determined using the algorithm described in Section 1. The original data for σ_y are converted to σ_θ using Equation (2.5). The figures show that the resampled data span the entire ranges of values provided by the experts. This is a significant advantage over other methods that were considered, which all employed least-square fitting of the original data.

Figures 2-18 and 2-19 compare the final results tabulated in Table 2-2 with the resampled data shown in Figures 2-14 through 2-17. In Figure 2-18, the comparisons are made at fixed quantile level. In Figure 2-19, the comparisons are made at fixed downwind distances. The agreement is remarkably good.

Table 2-2: Values for cross-wind and vertical dispersion coefficients for Pasquill-Gifford stability class C.

Quantile	Crosswind Dispersion Parameters		Vertical Dispersion Parameters	
	a (m)	b (dimensionless)	c (m)	d (dimensionless)
0.00	0.0631	0.865	0.0487	0.859
0.01	0.0963	0.865	0.0683	0.859
0.05	0.1564	0.865	0.0871	0.859
0.10	0.2000	0.865	0.1106	0.859
0.25	0.2805	0.865	0.1491	0.859
0.50	0.4063	0.865	0.2036	0.859
0.75	0.5939	0.865	0.3492	0.859
0.90	0.8257	0.865	0.5287	0.859
0.95	0.9735	0.865	0.7039	0.859
0.99	1.3720	0.865	1.2540	0.859
1.00	2.0763	0.865	1.8861	0.859
Mean	0.4031	0.865	0.2274	0.859
Mode	0.4379	0.865	0.1514	0.859

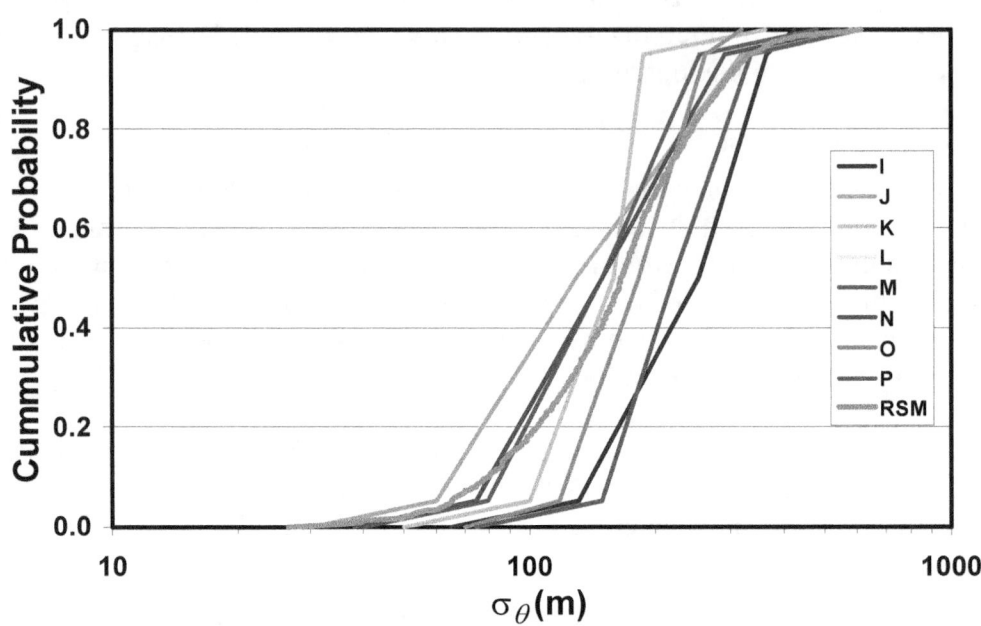

Figure 2-14. Expert and resampled data for σ_θ at 1 km downwind and Pasquill-Gifford stability class C

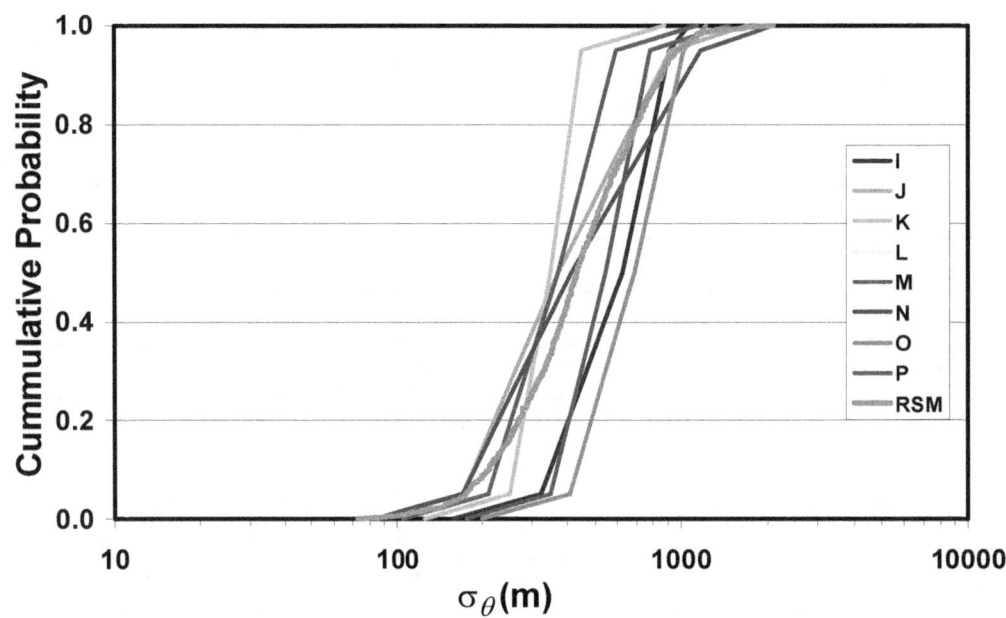

Figure 2-15. Expert and resampled data for σ_θ at 3 km downwind and Pasquill-Gifford stability class C

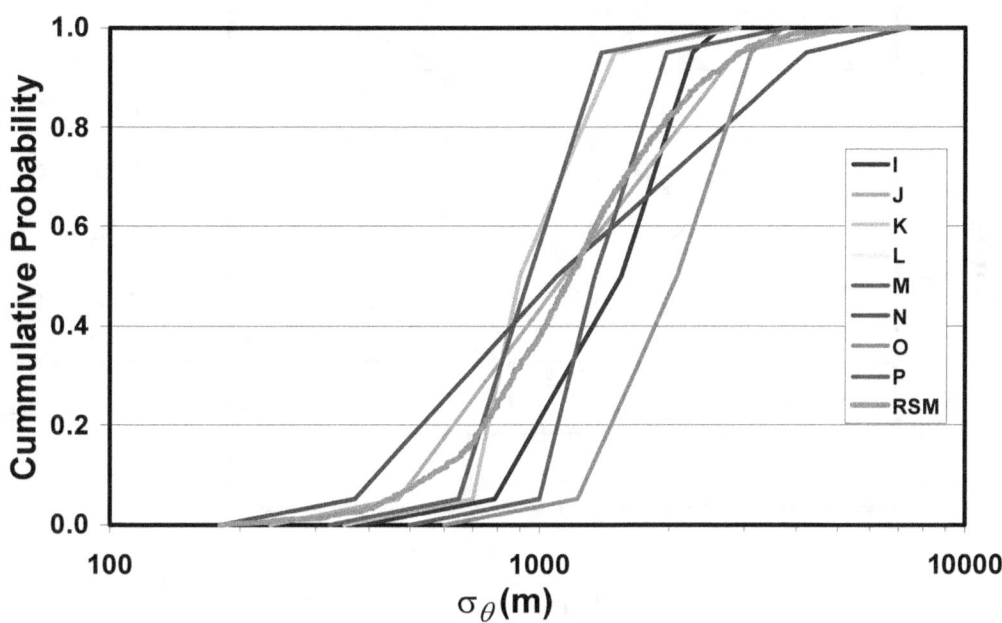

Figure 2-16. Expert and resampled data for σ_θ at 10 km downwind and Pasquill-Gifford stability class C

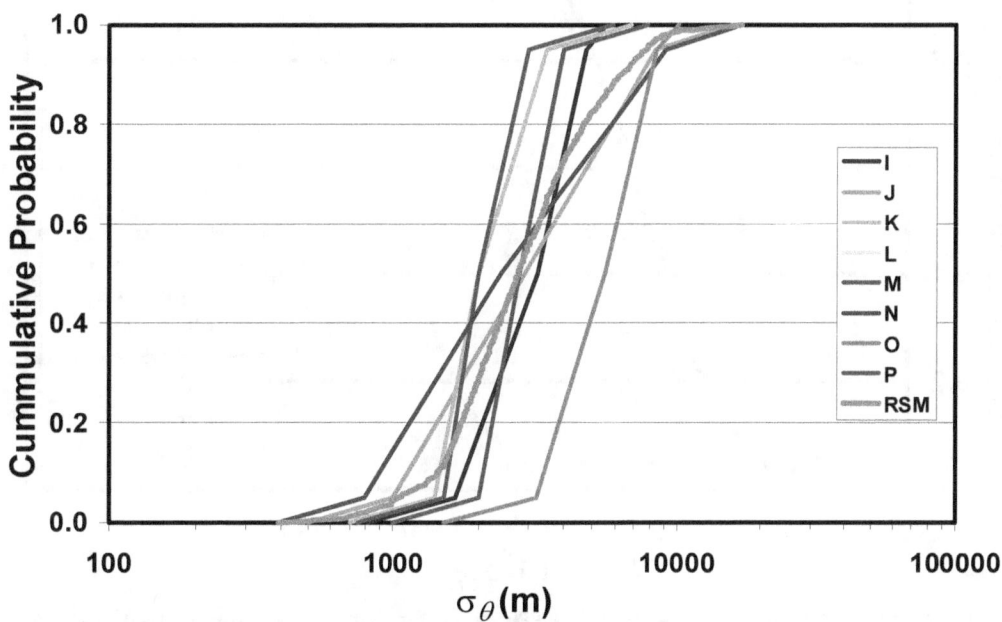

Figure 2-17. Expert and resampled data for σ_θ at 30 km downwind and Pasquill-Gifford stability class C

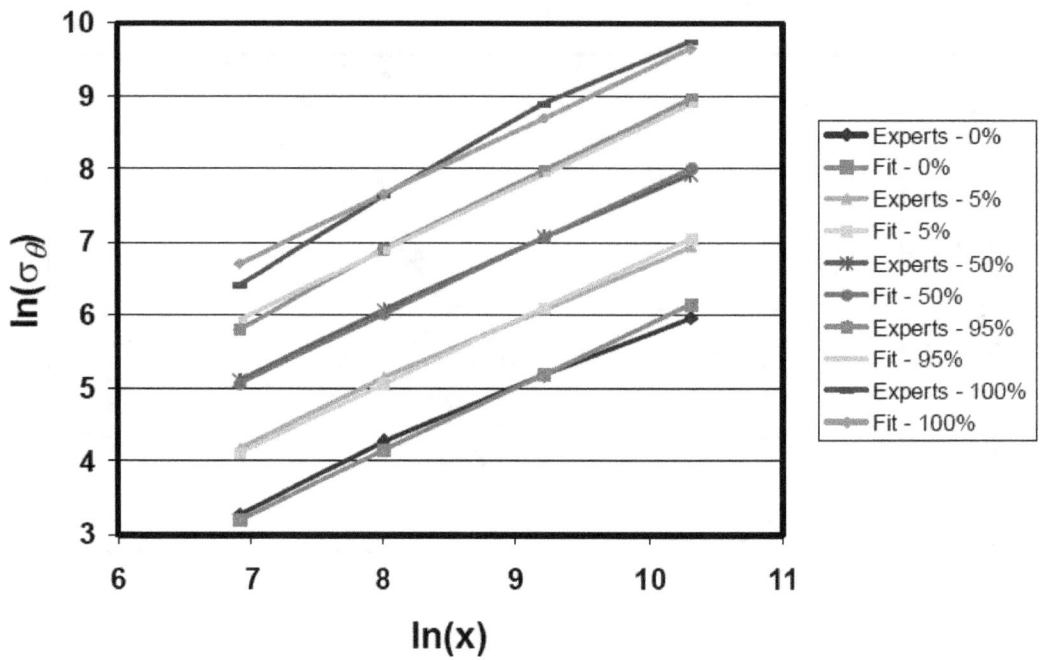

Figure 2-18.   Comparison of resampled expert data for σ_θ and Pasquill-Gifford stability class C with the calculated values using the parameters in Table 2-2

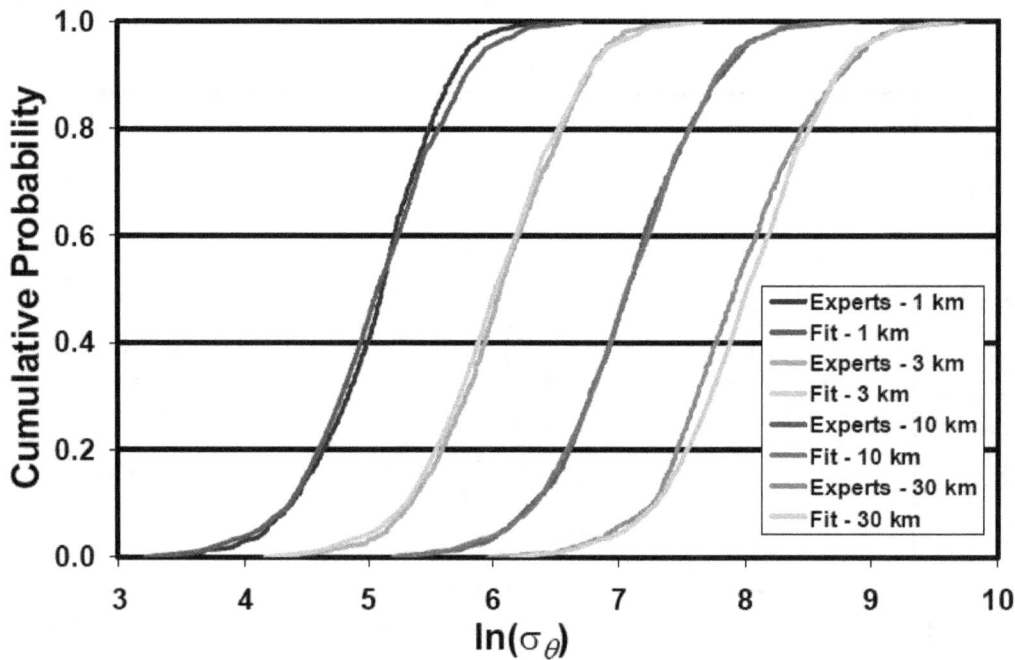

Figure 2-19.   Comparison of resampled expert data for σ_θ and Pasquill-Gifford stability class C with the calculated values using the parameters in Table 2-2

2.2.2 Vertical Dispersion

Figures 2-20 through 2-22 display the expert data for each of the eight experts and the resampled data that were created for this study. Missing values in the expert data are determined using the algorithm described in Section 1. The figures show that the resampled data span the entire ranges of values provided by the experts. This is a significant advantage over other methods that were considered, which all employed least-square fitting of the original data.

Figures 2-23 and 2-24 compare the final results tabulated in Table 2-2 with the resampled data shown in Figures 2-20 through 2-22. In Figure 2-23, the comparisons are made at fixed quantile level. In Figure 2-24, the comparisons are made at fixed downwind distances. The agreement is remarkably good.

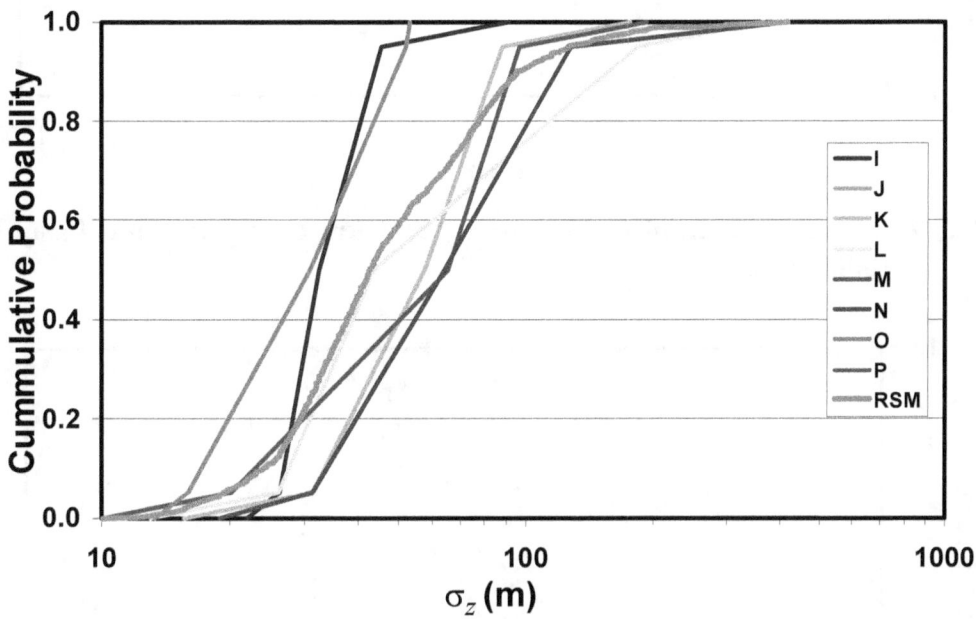

Figure 2-20. Expert and resampled data for σ_z at 0.5 km downwind and Pasquill-Gifford stability class C

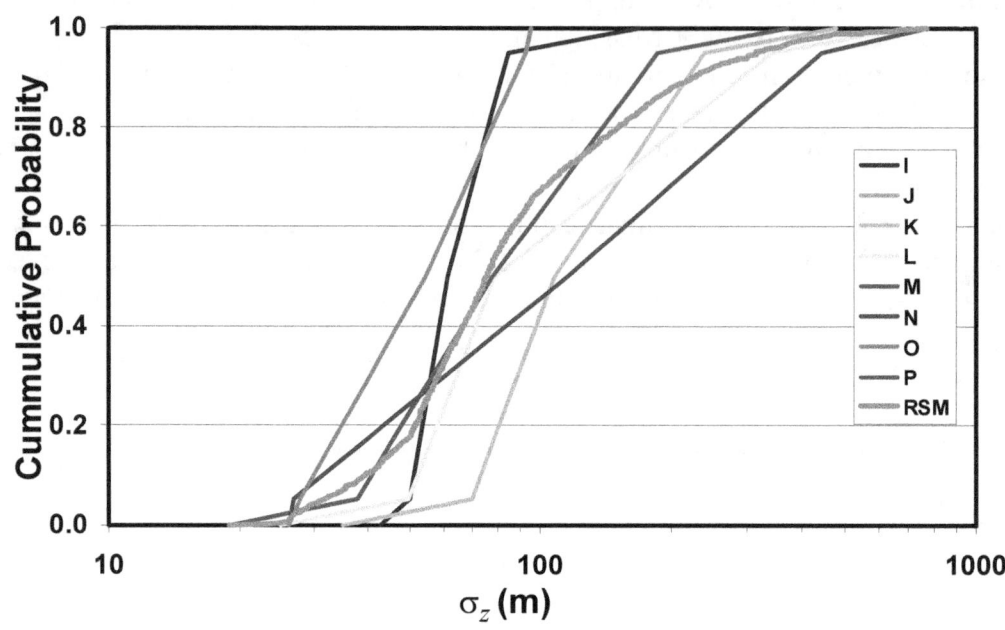

Figure 2-21. Expert and resampled data for σ_z at 1 km downwind and Pasquill-Gifford stability class C

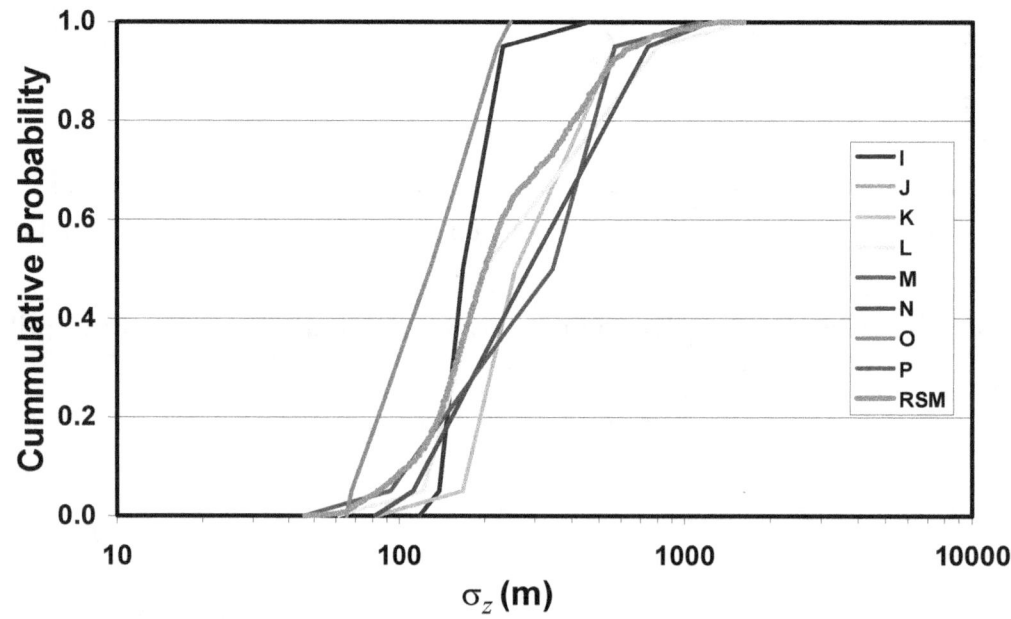

Figure 2-22. Expert and resampled data for σ_z at 3 km downwind and Pasquill-Gifford stability class C

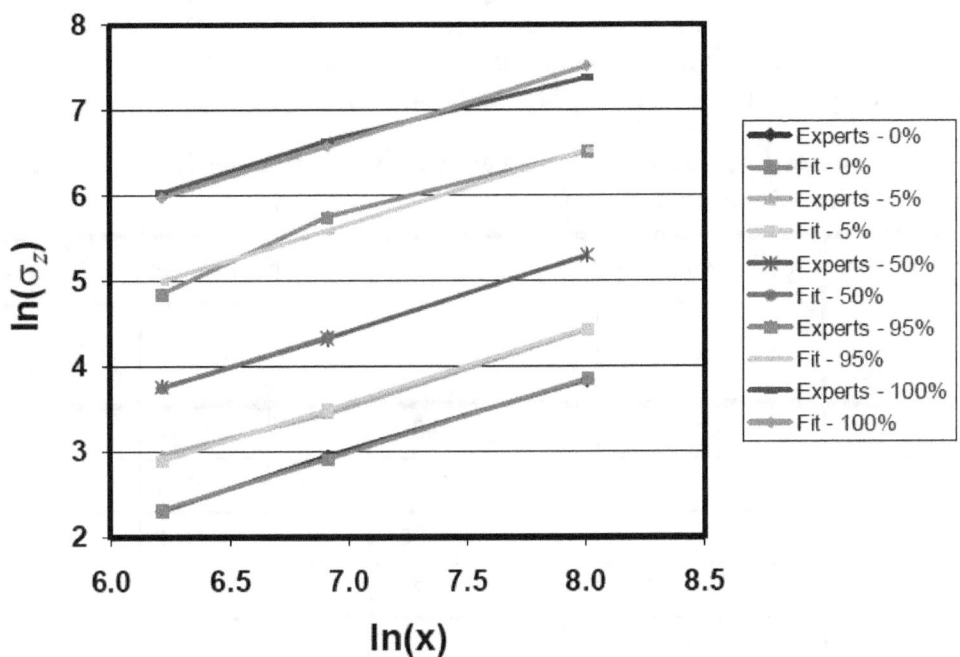

Figure 2-23. Comparison of resampled expert data for σ_z and Pasquill-Gifford stability class C with the calculated values using the parameters in Table 2-2

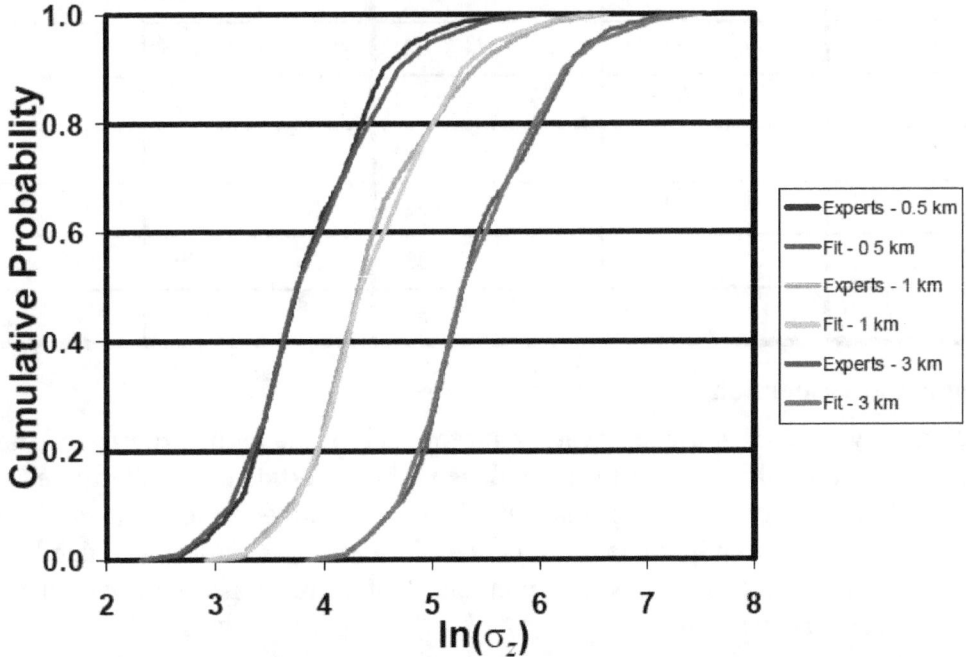

Figure 2-24. Comparison of resampled expert data for σ_z and Pasquill-Gifford stability class C with the calculated values using the parameters in Table 2-2

2.3 Distributions for Stability Class D

Table 2-3 provides values for a, b, c, and d from Equations (2.1) and (2.2) when the Pasquill-Gifford stability class is D. To interpolate values between the quantile levels provided in the table, it is preferable to interpolate the logarithms of a and c linearly.

Table 2-3: Values for cross-wind and vertical dispersion coefficients for Pasquill-Gifford stability class D.

Quantile	Crosswind Dispersion Parameters		Vertical Dispersion Parameters	
	a (m)	b (dimensionless)	c (m)	d (dimensionless)
0.00	0.0341	0.881	0.0421	0.751
0.01	0.0562	0.881	0.0752	0.751
0.05	0.0961	0.881	0.1161	0.751
0.10	0.1253	0.881	0.1404	0.751
0.25	0.1845	0.881	0.1821	0.751
0.50	0.2779	0.881	0.2636	0.751
0.75	0.4282	0.881	0.4224	0.751
0.90	0.6080	0.881	0.6048	0.751
0.95	0.7570	0.881	0.7504	0.751
0.99	1.1511	0.881	1.4634	0.751
1.00	1.7618	0.881	3.6880	0.751
Mean	0.2765	0.881	0.2796	0.751
Mode	0.2865	0.881	0.1842	0.751

2.3.1 Crosswind Dispersion

Figures 2-25 through 2-28 display the expert data for each of the eight experts and the resampled data that were created for this study. Missing values in the expert data are determined using the algorithm described in Section 1. The original data for σ_y are converted to σ_θ using Equation (2.5). The figures show that the resampled data span the entire ranges of values provided by the experts. This is a significant advantage over other methods that were considered, which all employed least-square fitting of the original data.

Figures 2-29 and 2-30 compare the final results tabulated in Table 2-3 with the resampled data shown in Figures 2-25 through 2-28. In Figure 2-29, the comparisons are made at fixed quantile level. In Figure 2-30, the comparisons are made at fixed downwind distances. The agreement is generally very good.

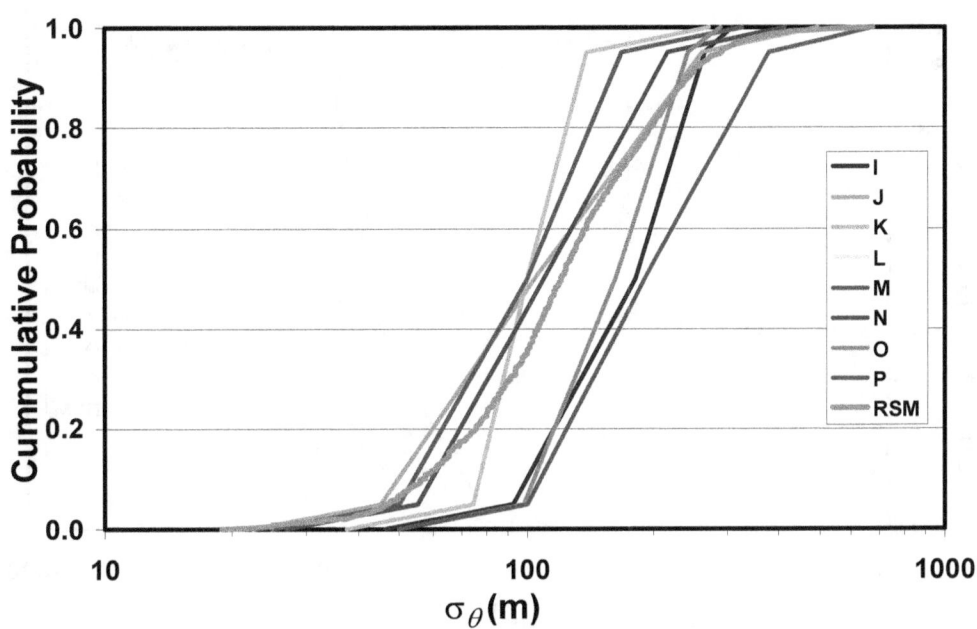

Figure 2-25. Expert and resampled data for σ_θ at 1 km downwind and Pasquill-Gifford stability class D

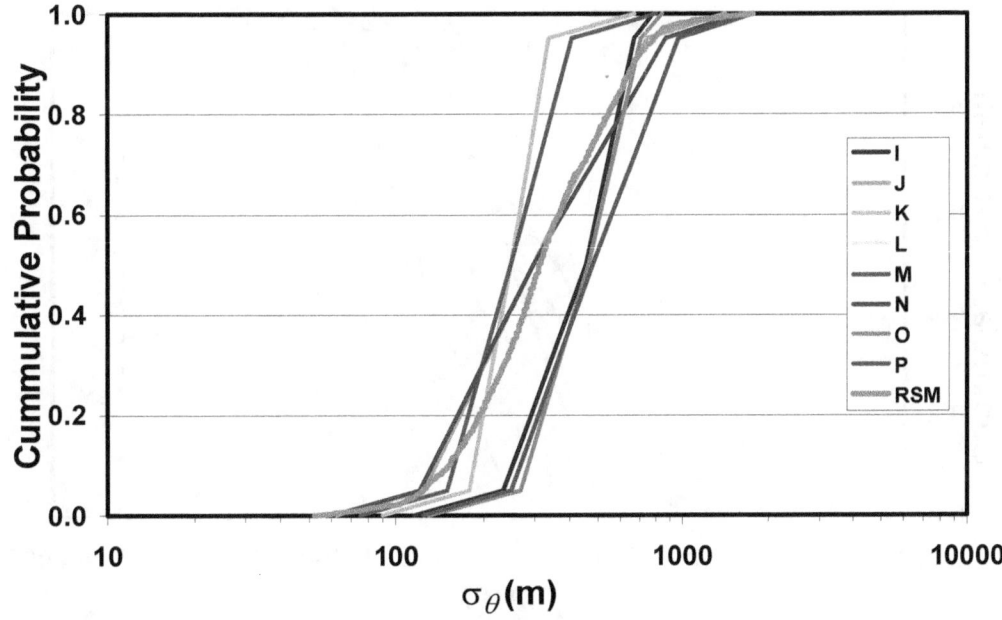

Figure 2-26. Expert and resampled data for σ_θ at 3 km downwind and Pasquill-Gifford stability class D

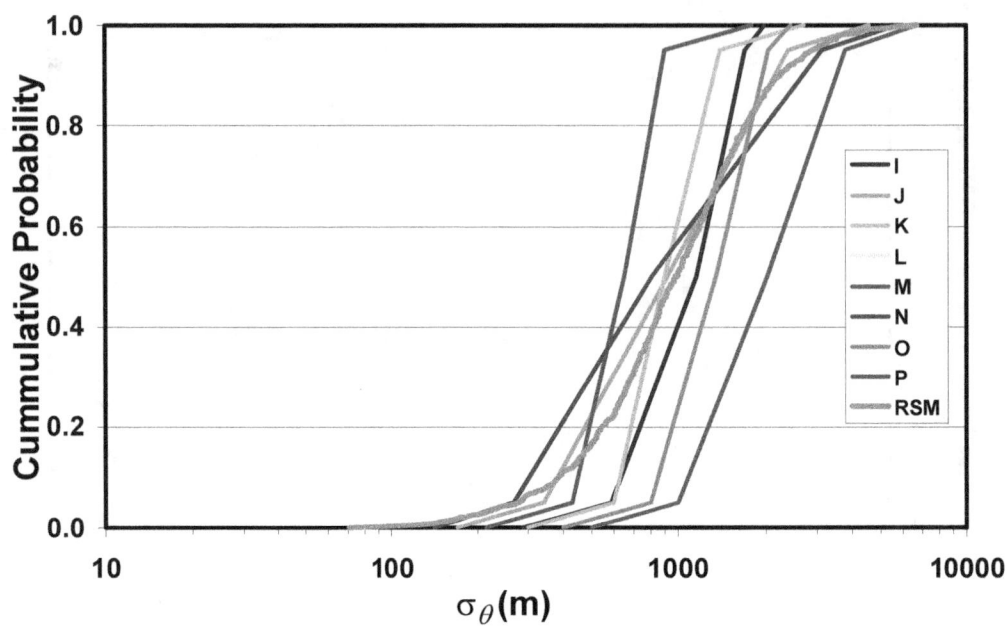

Figure 2-27. Expert and resampled data for σ_θ at 10 km downwind and Pasquill-Gifford stability class D

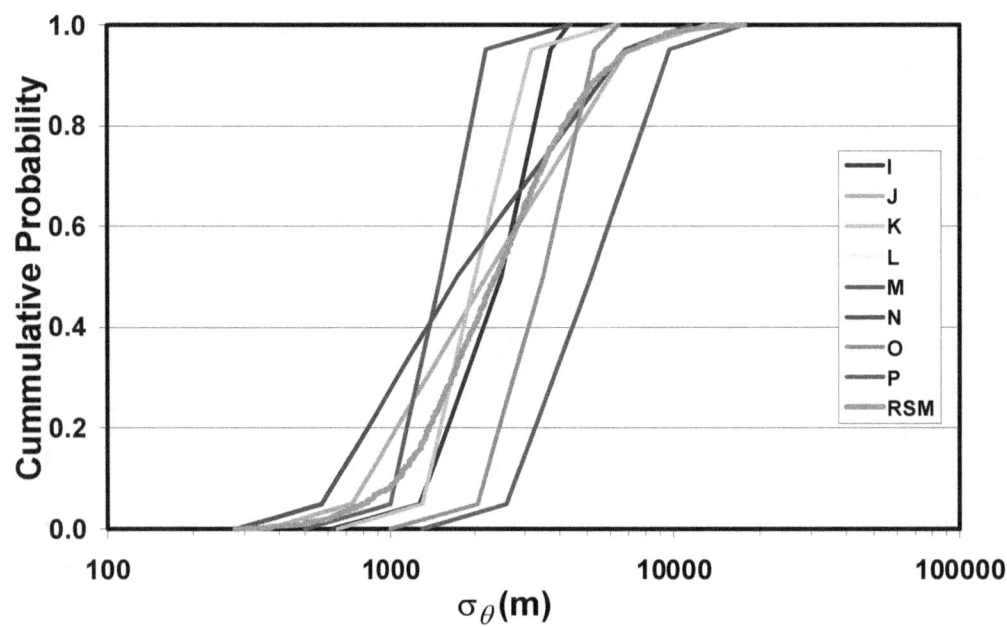

Figure 2-28. Expert and resampled data for σ_θ at 30 km downwind and Pasquill-Gifford stability class D

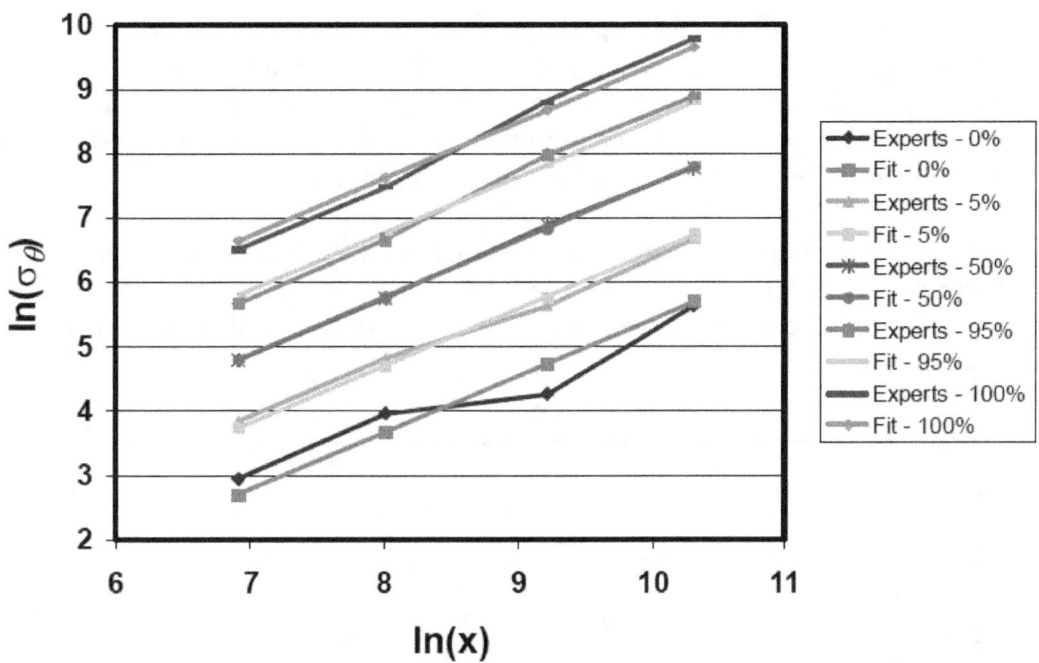

Figure 2-29. Comparison of resampled expert data for σ_θ and Pasquill-Gifford stability class D with the calculated values using the parameters in Table 2-3

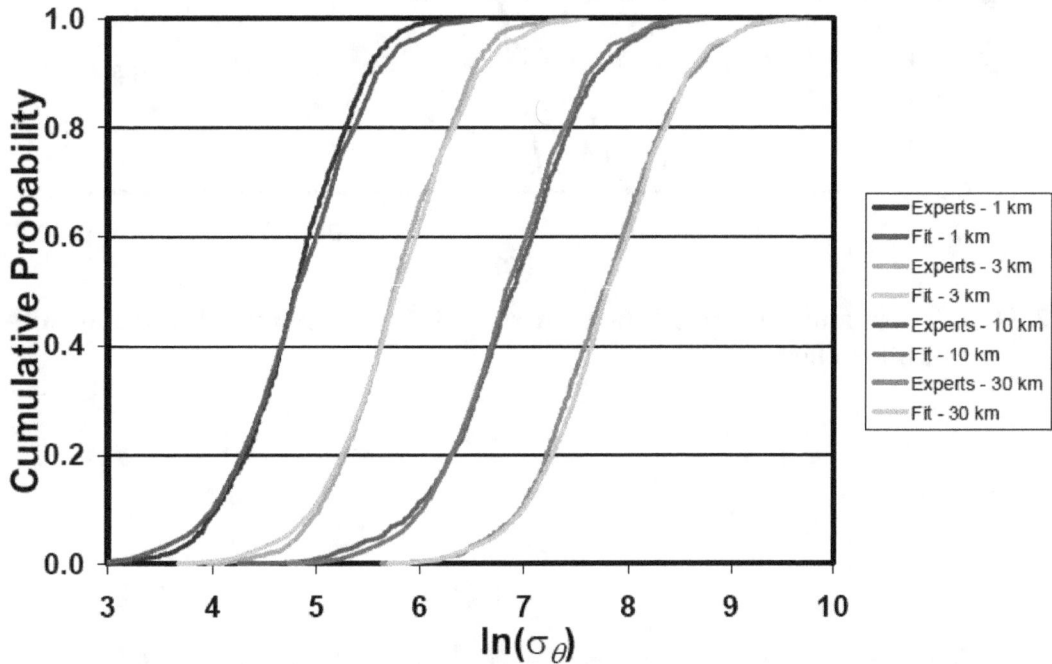

Figure 2-30. Comparison of resampled expert data for σ_θ and Pasquill-Gifford stability class D with the calculated values using the parameters in Table 2-3

2.3.2 Vertical Dispersion

Figures 2-31 through 2-33 display the expert data for each of the eight experts and the resampled data that were created for this study. Missing values in the expert data are determined using the algorithm described in Section 1. The figures show that the resampled data span the entire ranges of values provided by the experts. This is a significant advantage over other methods that were considered, which all employed least-square fitting of the original data.

Figures 2-34 and 2-35 compare the final results tabulated in Table 2-3 with the resampled data shown in Figures 2-31 through 2-33. In Figure 2-34, the comparisons are made at fixed quantile level. In Figure 2-35, the comparisons are made at fixed downwind distances. The agreement is remarkably good.

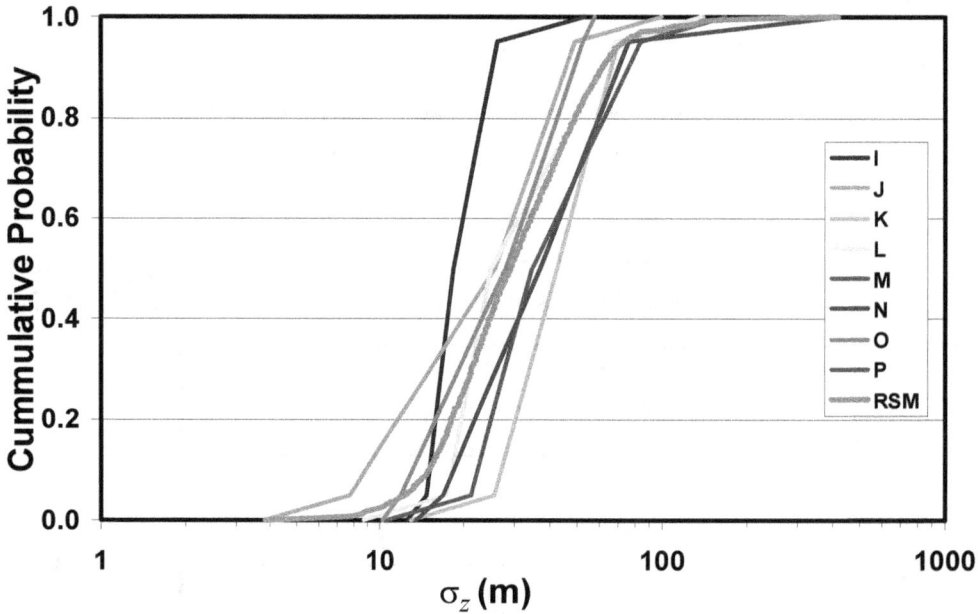

Figure 2-31. **Expert and resampled data for σ_z at 0.5 km downwind and Pasquill-Gifford stability class D**

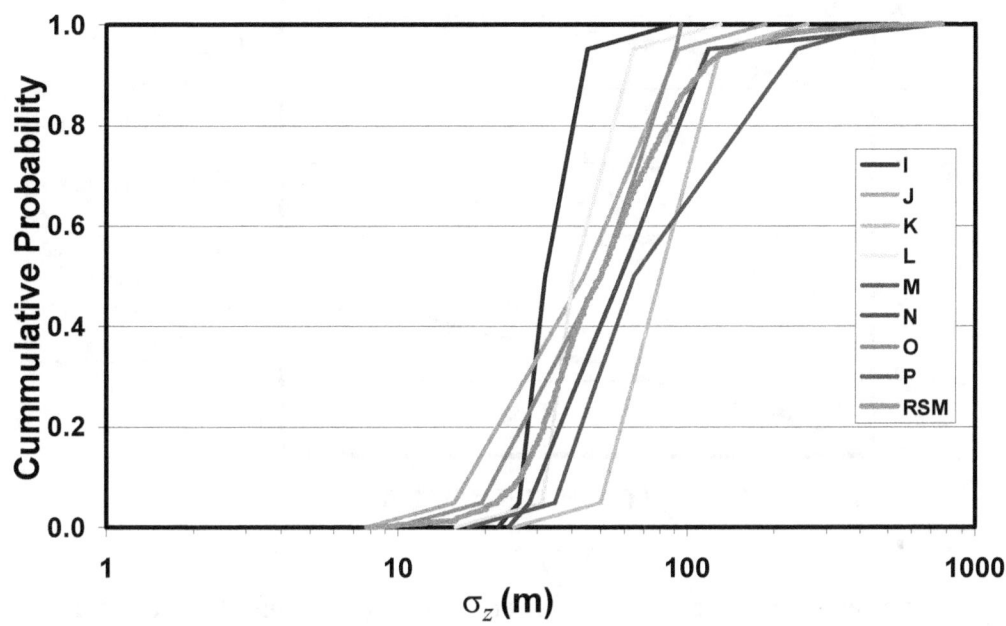

Figure 2-32. Expert and resampled data for σ_z at 1 km downwind and Pasquill-Gifford stability class D

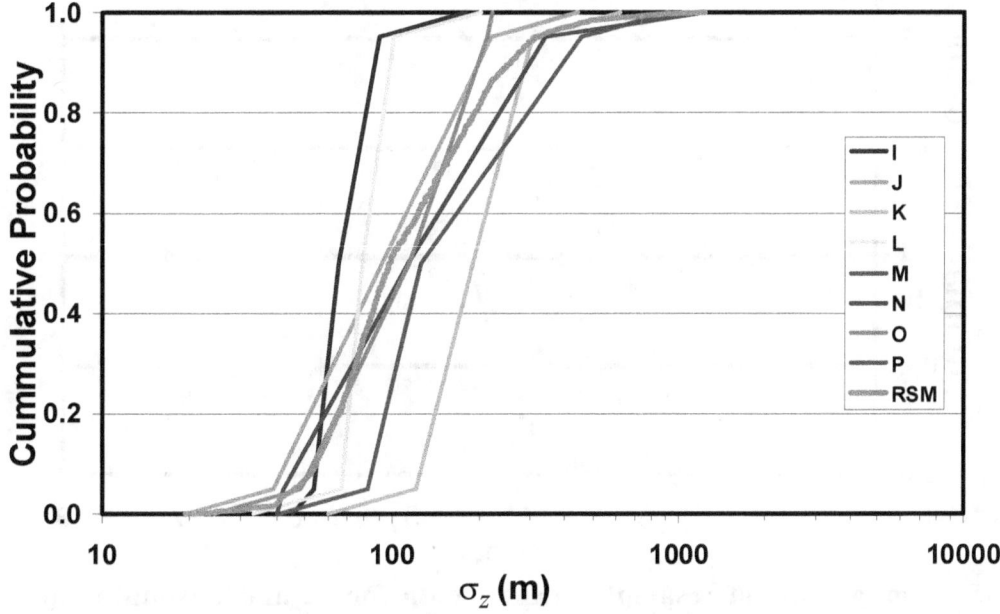

Figure 2-33. Expert and resampled data for σ_z at 3 km downwind and Pasquill-Gifford stability class D

31

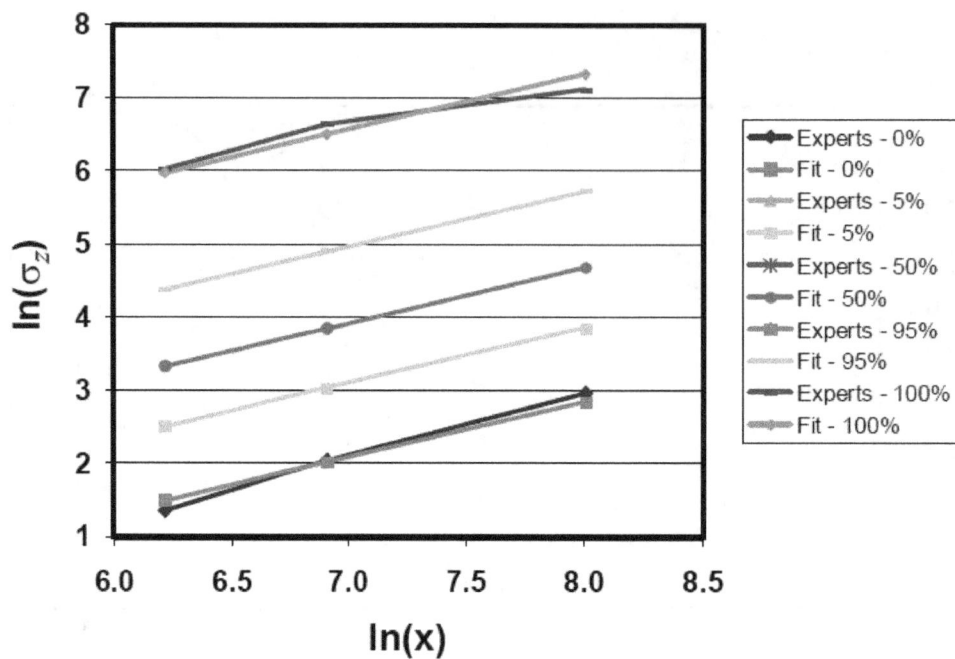

Figure 2-34. Comparison of resampled expert data for σ_z and Pasquill-Gifford stability class D with the calculated values using the parameters in Table 2-3

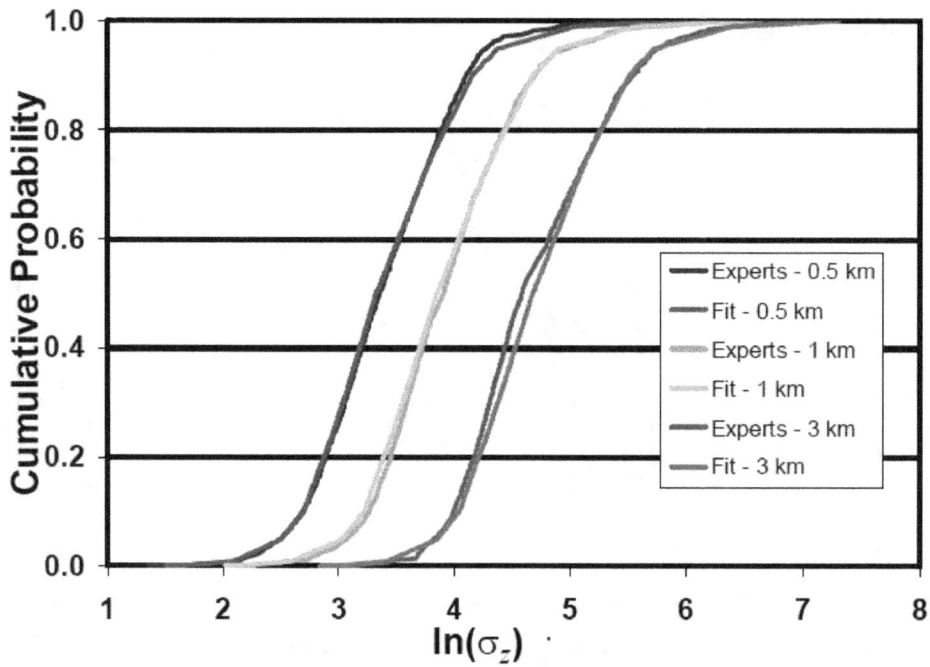

Figure 2-35. Comparison of resampled expert data for σ_z and Pasquill-Gifford stability class D with the calculated values using the parameters in Table 2-3

2.4 Distributions for Stability Class E/F

Table 2-4 provides values for a, b, c, and d from Equations (2.1) and (2.2) when the Pasquill-Gifford stability class is E or F. To interpolate values between the quantile levels provided in the table, it is preferable to interpolate the logarithms of a and c linearly.

Table 2-4: Values for cross-wind and vertical dispersion coefficients for Pasquill-Gifford stability class E/F.

Quantile	Crosswind Dispersion Parameters		Vertical Dispersion Parameters	
	a (m)	b (dimensionless)	c (m)	d (dimensionless)
0.00	0.0212	0.866	0.0533	0.619
0.01	0.0376	0.866	0.0756	0.619
0.05	0.0575	0.866	0.1141	0.619
0.10	0.0768	0.866	0.1310	0.619
0.25	0.1193	0.866	0.1598	0.619
0.50	0.2158	0.866	0.2463	0.619
0.75	0.3730	0.866	0.4617	0.619
0.90	0.5458	0.866	0.8180	0.619
0.95	0.6583	0.866	1.1260	0.619
0.99	0.9467	0.866	2.2051	0.619
1.00	1.5307	0.866	4.5386	0.619
Mean	0.2081	0.866	0.2874	0.619
Mode	0.2169	0.866	0.1458	0.619

2.4.1 Crosswind Dispersion

Figures 2-36 through 2-39 display the expert data for each of the eight experts and the resampled data that were created for this study. Missing values in the expert data are determined using the algorithm described in Section 1. The original data for σ_y are converted to σ_θ using Equation (2.5). The figures show that the resampled data span the entire ranges of values provided by the experts. This is a significant advantage over other methods that were considered, which all employed least-square fitting of the original data.

Figures 2-40 and 2-41 compare the final results tabulated in Table 2-4 with the resampled data shown in Figures 2-36 through 2-39. In Figure 2-40, the comparisons are made at fixed quantile level. In Figure 2-41, the comparisons are made at fixed downwind distances. The agreement is remarkably good.

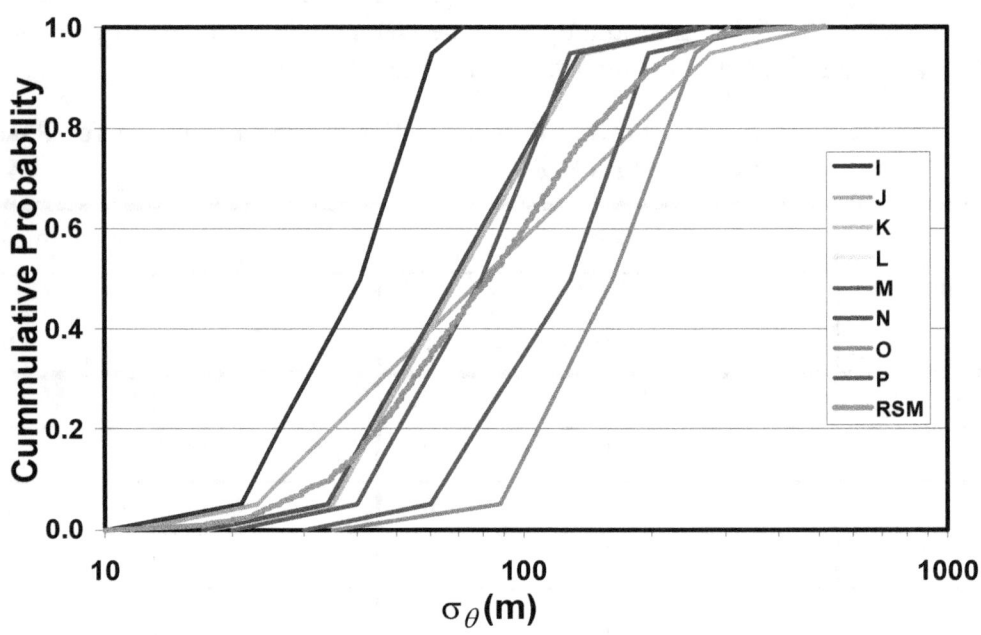

Figure 2-36. Expert and resampled data for σ_θ at 1 km downwind and Pasquill-Gifford stability class E/F

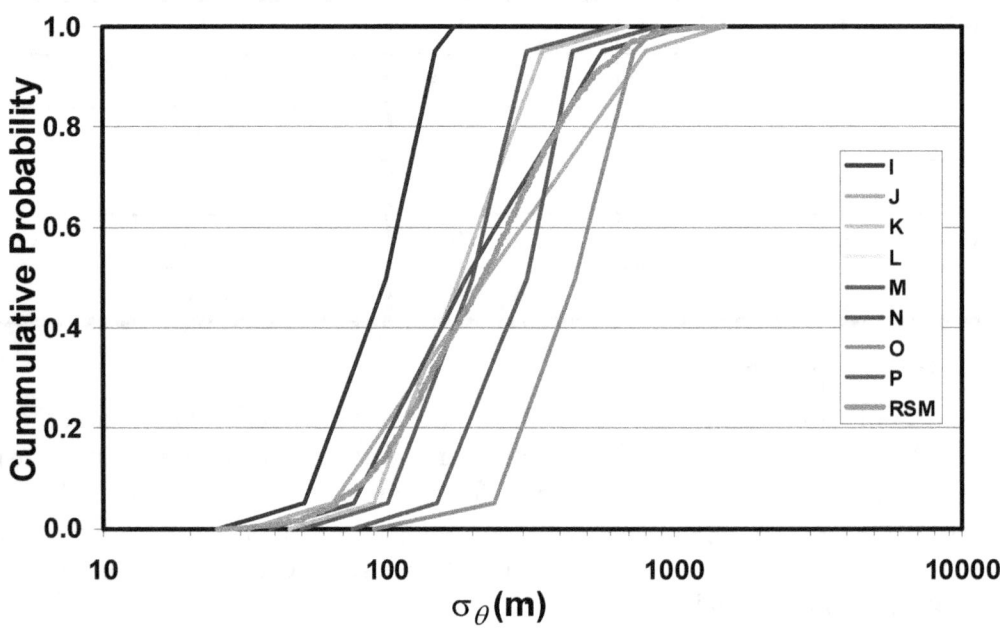

Figure 2-37. Expert and resampled data for σ_θ at 3 km downwind and Pasquill-Gifford stability class E/F

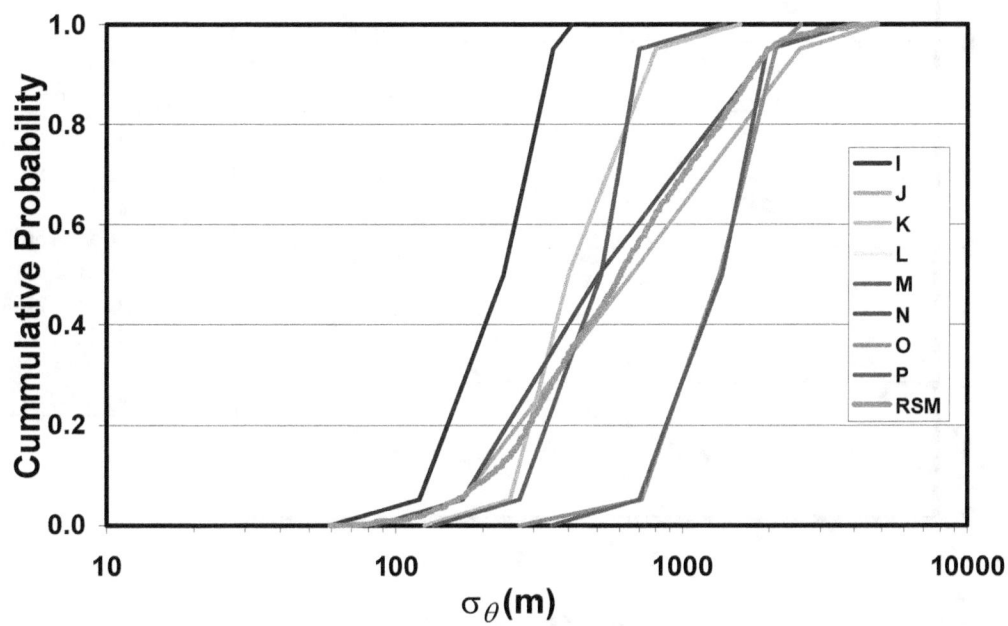

Figure 2-38. Expert and resampled data for σ_θ at 10 km downwind and Pasquill-Gifford stability class E/F

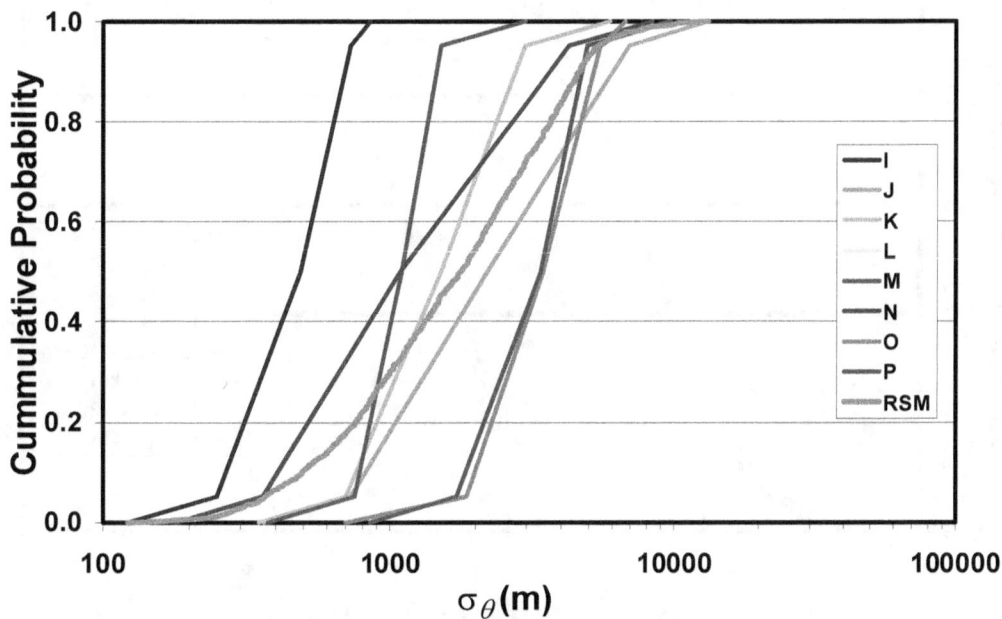

Figure 2-39. Expert and resampled data for σ_θ at 30 km downwind and Pasquill-Gifford stability class E/F

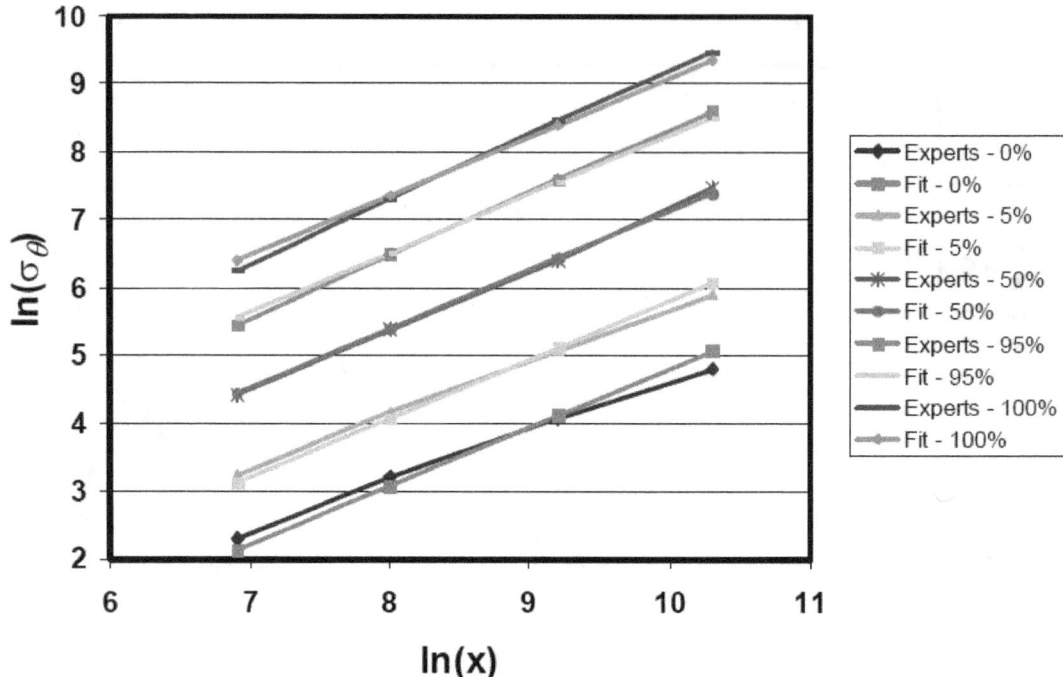

Figure 2-40. Comparison of resampled expert data for σ_θ and Pasquill-Gifford stability class E/F with the calculated values using the parameters in Table 2-4

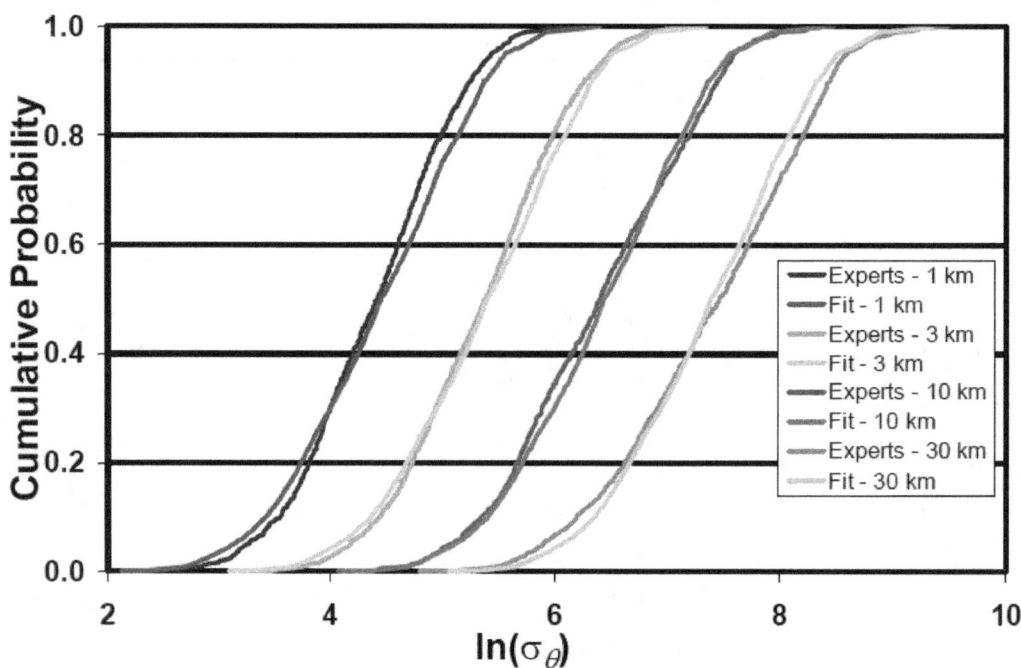

Figure 2-41. Comparison of resampled expert data for σ_θ and Pasquill-Gifford stability class E/F with the calculated values using the parameters in Table 2-4

2.4.2 Vertical Dispersion

Figures 2-42 through 2-44 display the expert data for each of the eight experts and the resampled data that were created for this study. Missing values in the expert data are determined using the algorithm described in Section 1. The figures show that the resampled data span the entire ranges of values provided by the experts. This is a significant advantage over other methods that were considered, which all employed least-square fitting of the original data.

Figures 2-45 and 2-46 compare the final results tabulated in Table 2-4 with the resampled data shown in Figures 2-42 through 2-44. In Figure 2-45, the comparisons are made at fixed quantile level. In Figure 2-46, the comparisons are made at fixed downwind distances. The agreement is generally very good.

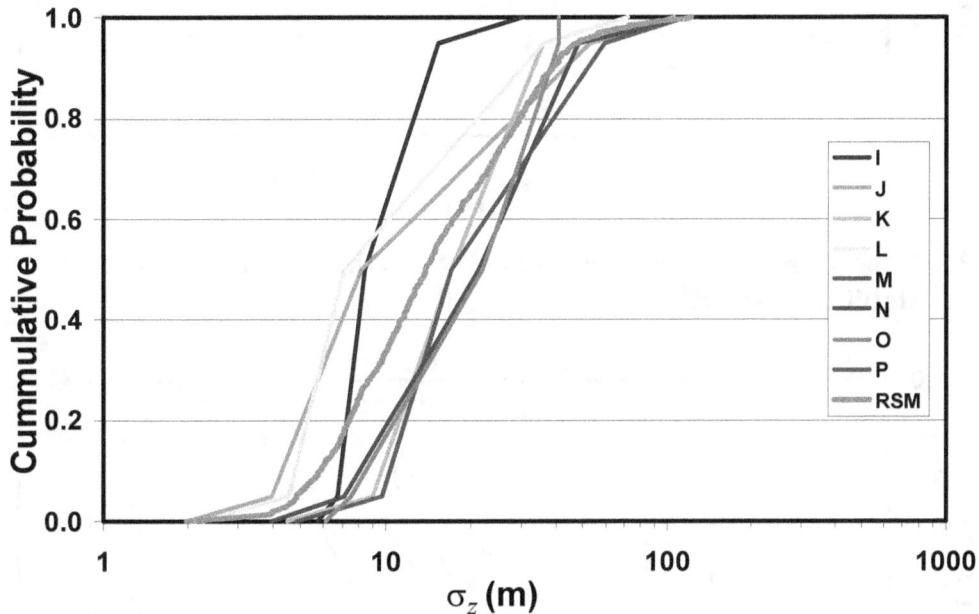

Figure 2-42. **Expert and resampled data for σ_z at 0.5 km downwind and Pasquill-Gifford stability class E/F**

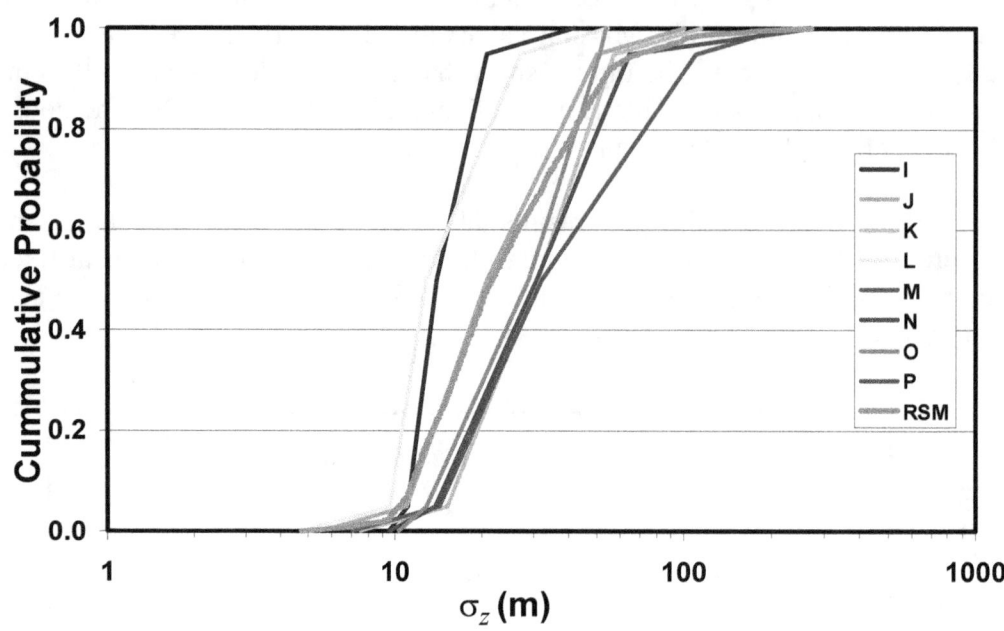

Figure 2-43. Expert and resampled data for σ_z at 1 km downwind and Pasquill-Gifford stability class E/F

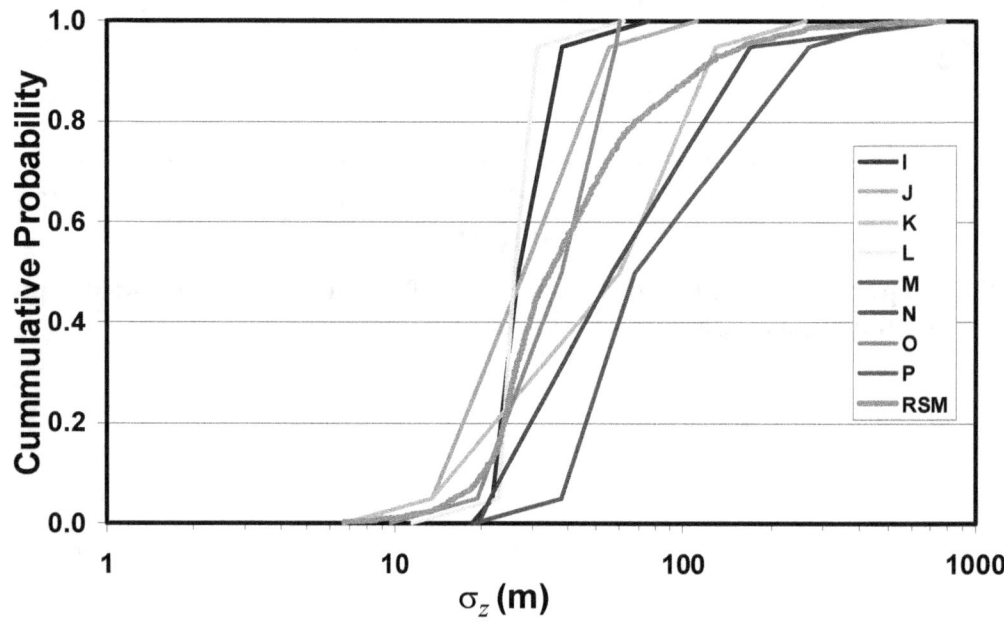

Figure 2-44. Expert and resampled data for σ_z at 3 km downwind and Pasquill-Gifford stability class E/F

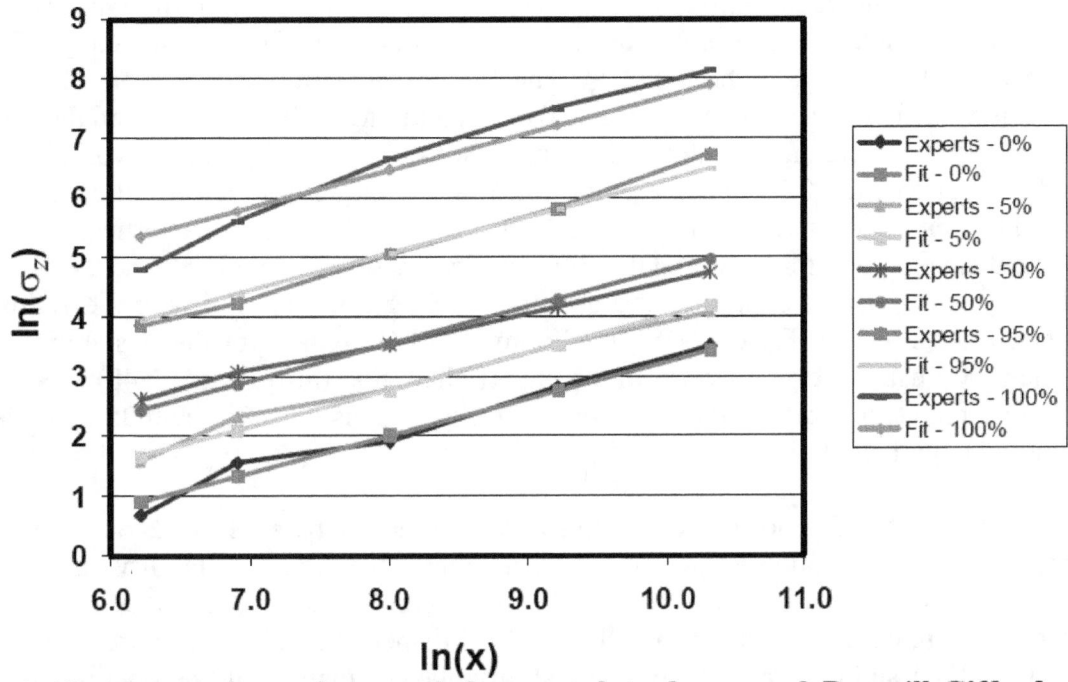

Figure 2-45. Comparison of resampled expert data for σ_z and Pasquill-Gifford stability class E/F with the calculated values using the parameters in Table 2-4

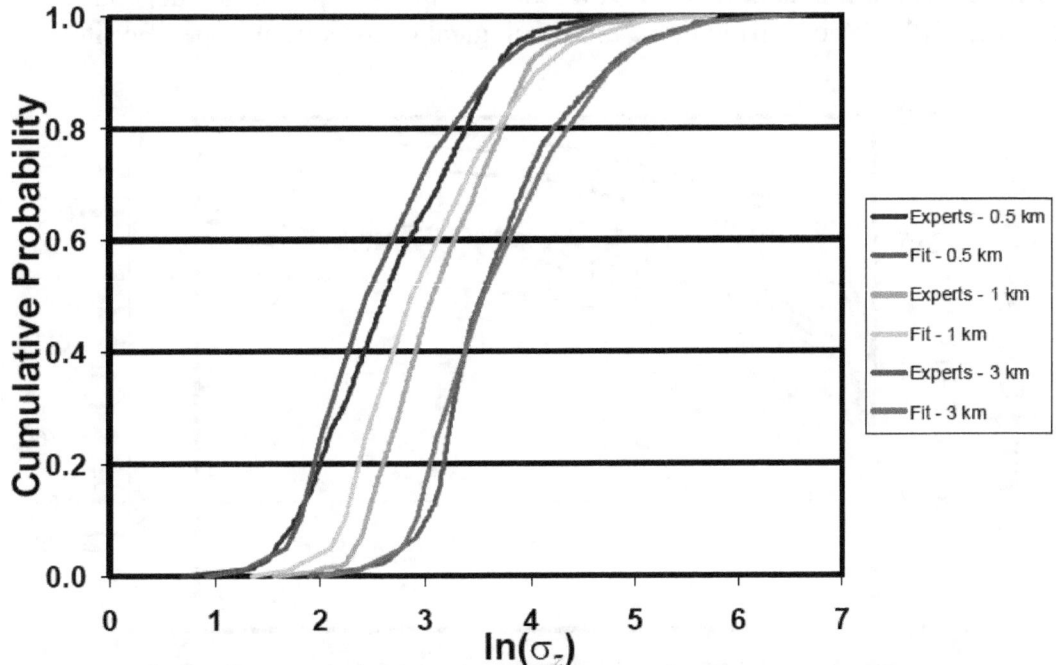

Figure 2-46. Comparison of resampled expert data for σ_z and Pasquill-Gifford stability class E/F with the calculated values using the parameters in Table 2-4

2.5   Comparisons with Tadmor and Gur Data

The data reported in Tables 2-1 through 2-4 are compared with the Tadmor and Gur (T&G) data used in one of the standard sample problems distributed with the MACCS2 code, Sample Problem A [Reference 7], in Figures 2-47 through 2-54. The Tadmor and Gur representation for dispersion has been widely used in reactor consequence analyses, including NUREG-1150, and therefore has both current and historical significance. The comparisons are generally quite favorable. For the crosswind dispersions, the 0.50 quantile curve from this work agrees well in all cases with the Tadmor and Gur curve or curves (Figures 2-47, 2-49, 2-51, and 2-53). For the vertical dispersions, the agreement is very good for the C and the E/F stability classes (Figures 2-50 and 2-54). The expert data for vertical dispersion at the 0.50 quantile level agrees very well with the Tadmor and Gur B stability class curve (Figure 2-48); the Tadmor and Gur curve for the A stability class is close to the 1.00 quantile level curve from the expert elicitation (also Figure 2-48). The Tadmor and Gur curve for vertical dispersion and the D stability class is close to the 0.05 quantile level curve from the expert elicitation (Figure 2-52).

In all cases but one, the Tadmor and Gur curves are very nearly parallel to the ones obtained in this study. The single exception is for vertical dispersion in the A and B stability classes, where the Tadmor and Gur curve for stability class A is significantly steeper than those obtained from the expert data (Figure 2-48). This means that vertical dispersion can occur more rapidly under the original Tadmor and Gur formulation for stability class A than would be obtained using the combined stability class A and B from this work, even at the 1.00 quantile level. The significance of this difference for a calculation in which weather sampling is performed depends on the probability of stability class A occurring in the meteorological conditions at a specific site.

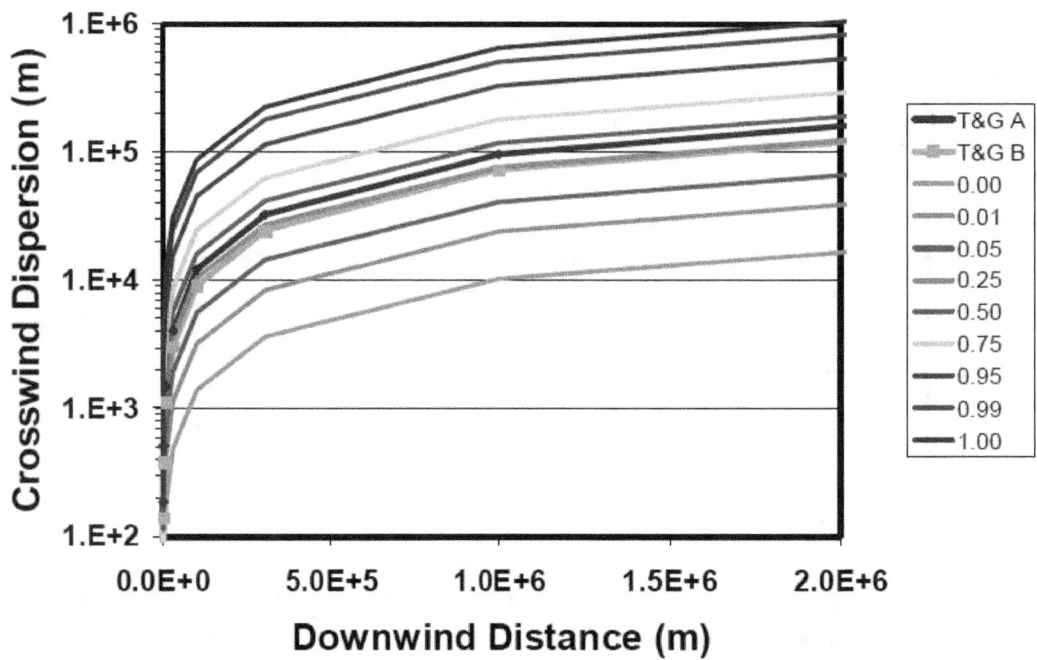

Figure 2-47.   Comparison between the Tadmor and Gur crosswind dispersion data and the processed expert data for the eleven quantile values given in Table 2-1 for stability classes A and B

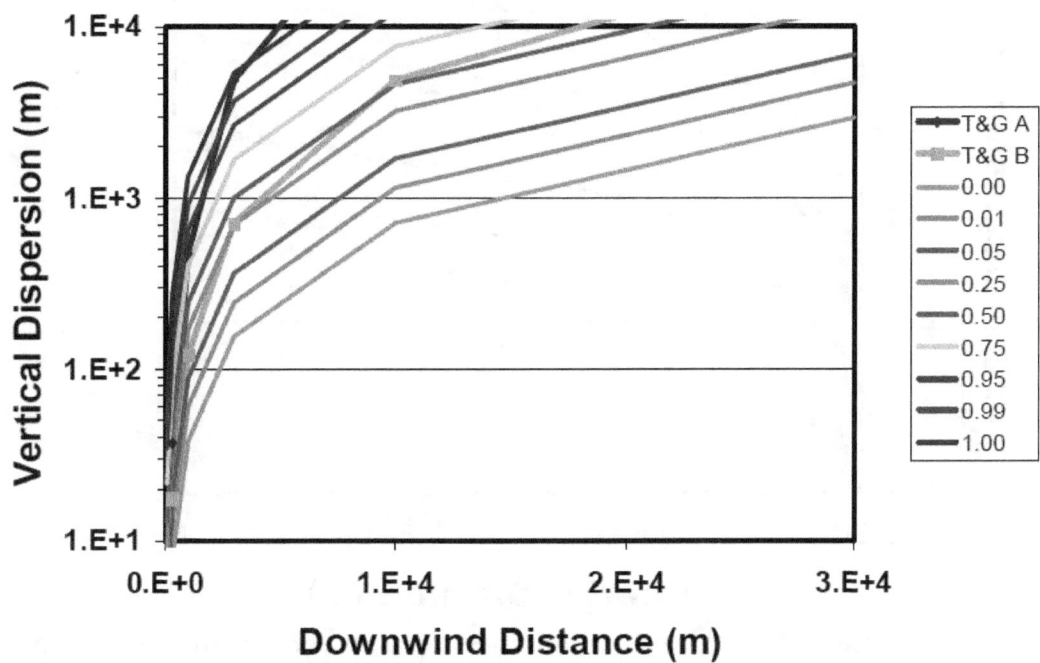

Figure 2-48. Comparison between the Tadmor and Gur vertical dispersion data and the processed expert data for the eleven quantile values given in Table 2-1 for stability classes A and B

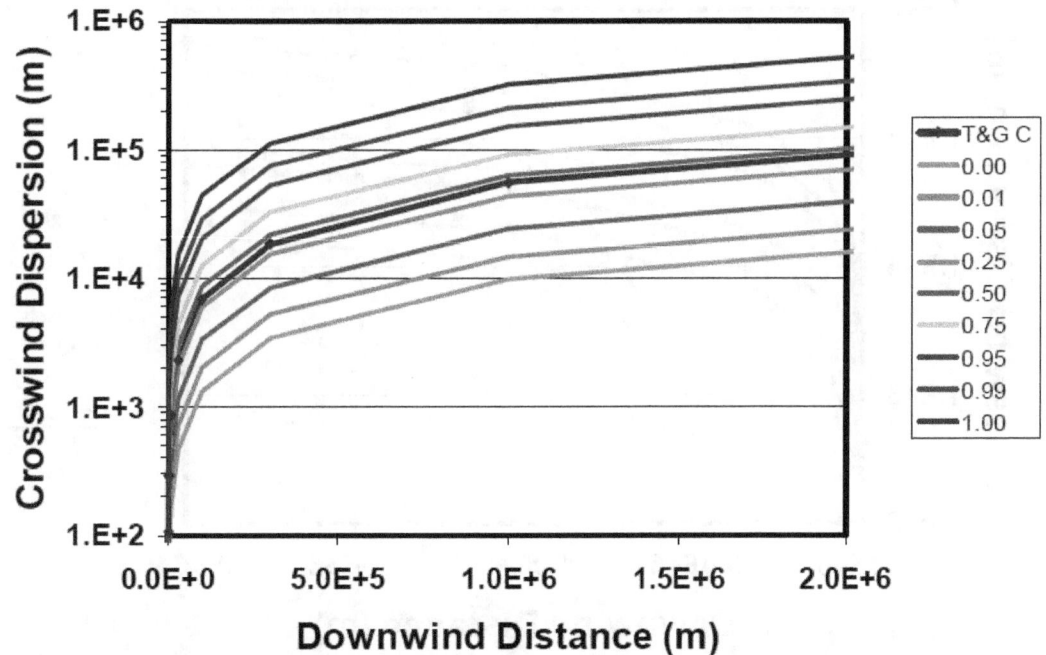

Figure 2-49. Comparison between the Tadmor and Gur crosswind dispersion data and the processed expert data for the eleven quantile values given in Table 2-2 for stability class C

41

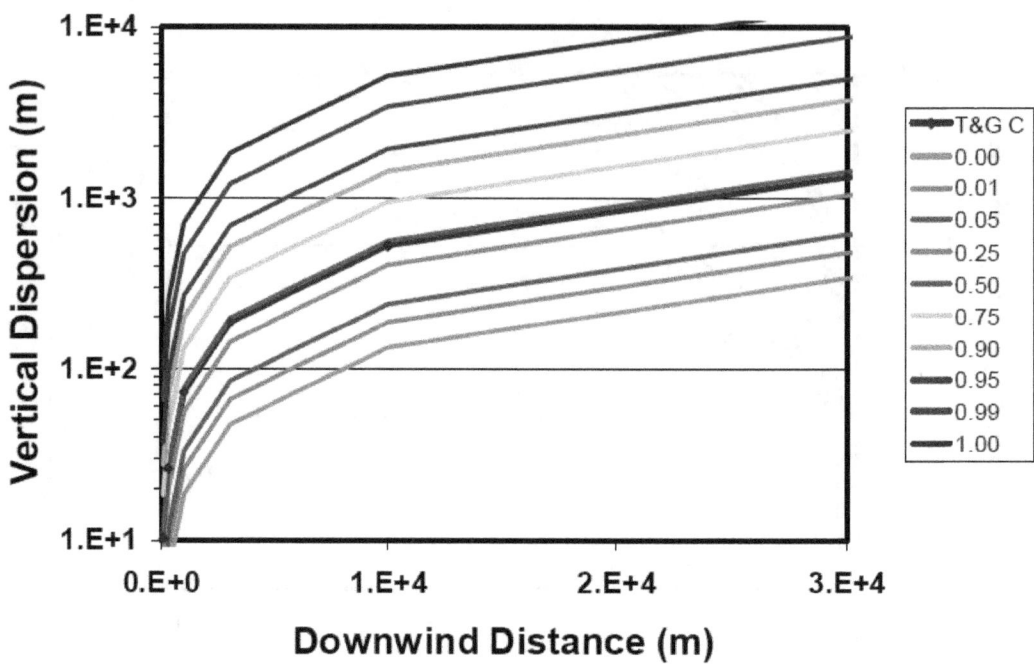

Figure 2-50. Comparison between the Tadmor and Gur vertical dispersion data and the processed expert data for the eleven quantile values given in Table 2-2 for stability class C

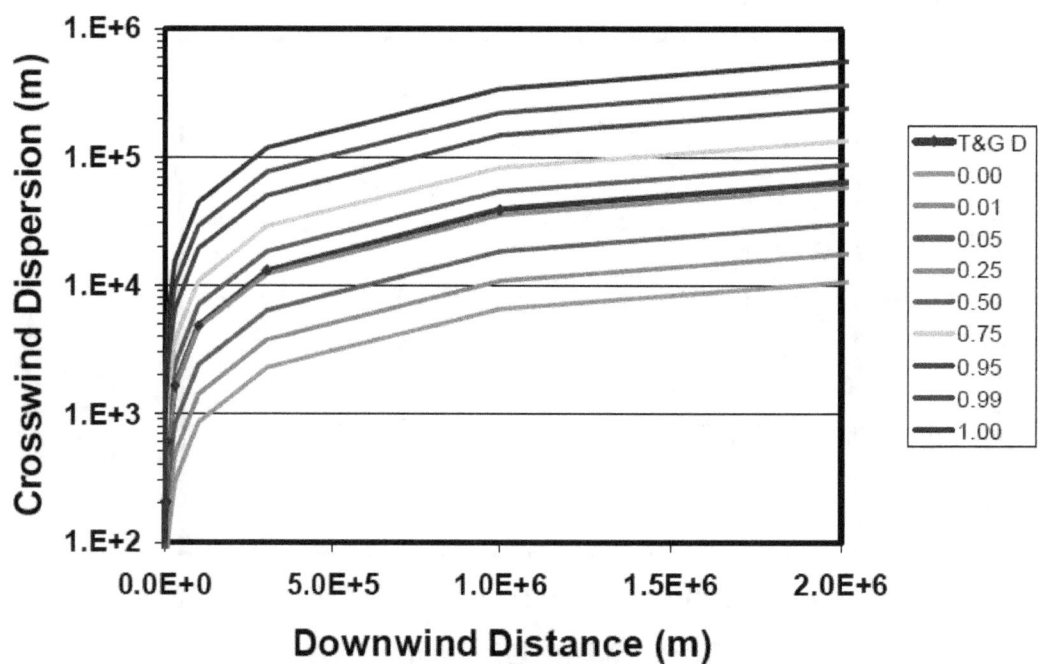

Figure 2-51. Comparison between the Tadmor and Gur crosswind dispersion data and the processed expert data for the eleven quantile values given in Table 2-3 for stability class D

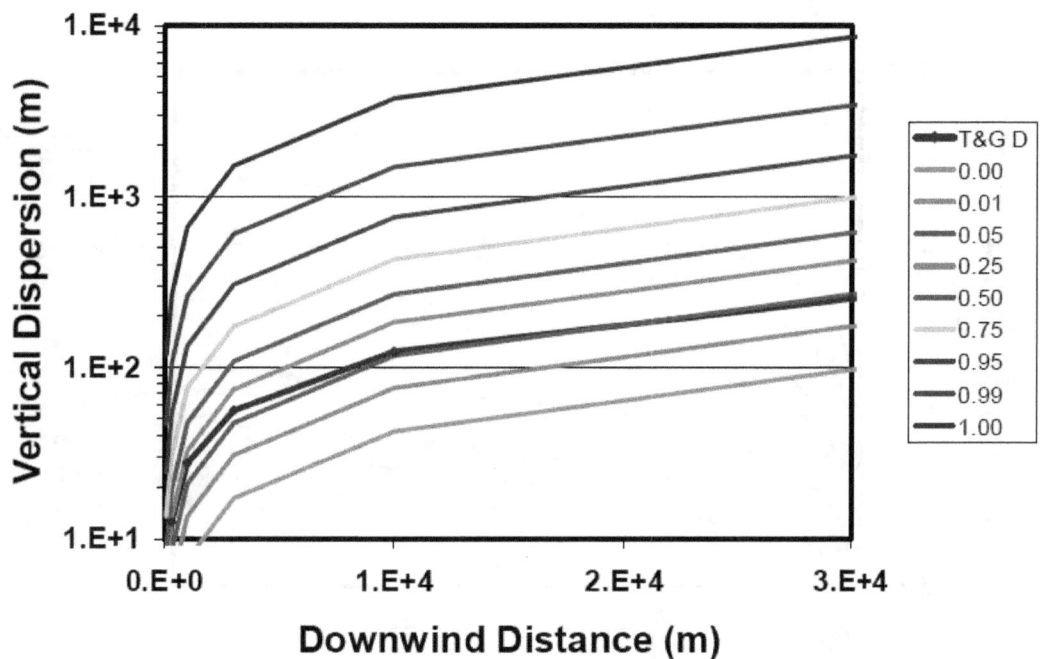

Figure 2-52. Comparison between the Tadmor and Gur vertical dispersion data and the processed expert data for the eleven quantile values given in Table 2-3 for stability class D

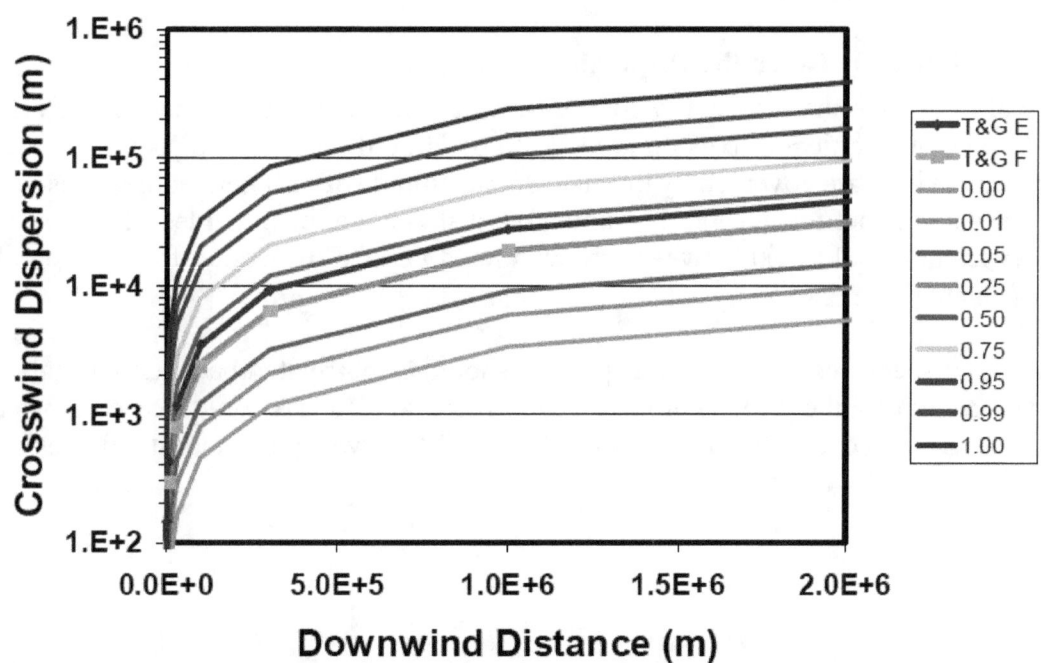

Figure 2-53. Comparison between the Tadmor and Gur crosswind dispersion data and the processed expert data for the eleven quantile values given in Table 2-4 for stability classes E and F

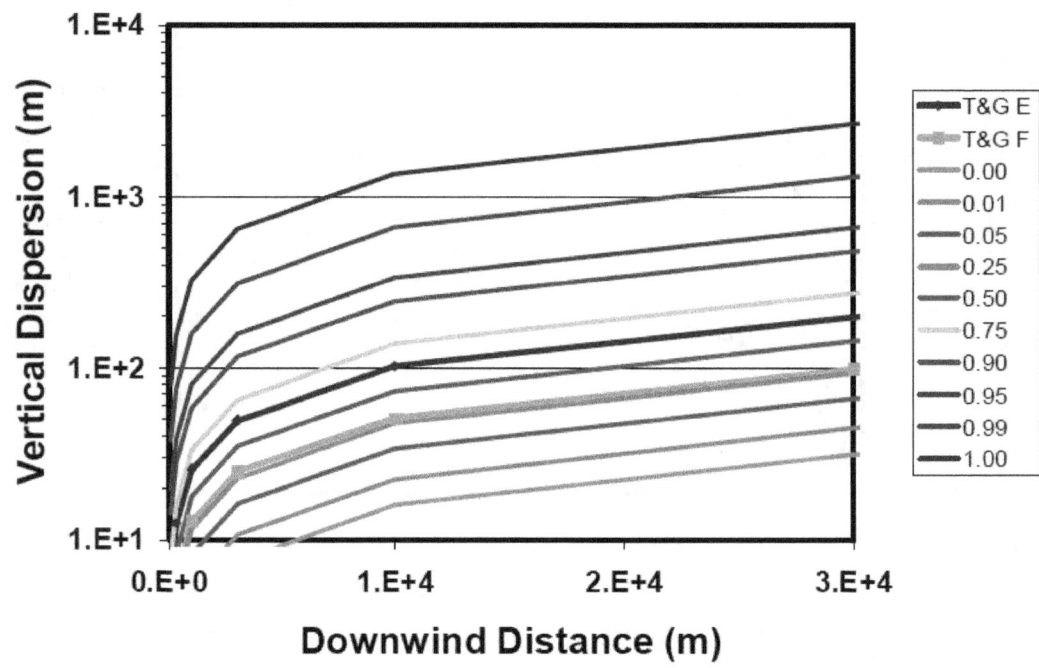

Figure 2-54. Comparison between the Tadmor and Gur vertical dispersion data and the processed expert data for the eleven quantile values given in Table 2-4 for stability classes E and F

2.6 Correlations Between the Dispersion Data

It is logical to assume that crosswind and vertical dispersion are correlated for a given stability class. For example, vertical dispersion is likely to be large if crosswind dispersion is determined to be large since both are governed by the size and magnitude of turbulent eddies. It is not obvious that the correlation coefficient should be unity, but it should not be much less than unity. A correlation coefficient of 0.9 to 1.0 between crosswind and vertical dispersion for each stability class seems reasonable.

Furthermore, it seems reasonable that dispersion should be correlated between the stability classes. If this were not so, the dispersion values for a more stable class could exceed those for a less stable class. Again, a correlation coefficient of 0.9 to 1.0 between the stability classes seems reasonable.

3.0 DRY DEPOSITION

Dry deposition velocities in MACCS2 are treated as being dependent on aerosol size, but as being independent of wind conditions and surface roughness. In the expert solicitation, both wind speed and surface roughness were considered as variables. The experts provided data for two wind speeds, 2 and 5 m/s, and for three surface roughnesses, corresponding to prairie, forest, and urban terrains.

The approach used in this work is to retain important dependencies so that the user can decide what is most appropriate for their specific situation. The final results for dry deposition velocities account for particle size, wind speed, surface roughness, and degree of belief. Because only particle size can be accounted for in the current MACCS2 dry deposition model, the user is required to choose representative values of wind speed, surface roughness, and quantile. Evaluation of the expert results indicates that quantile is significantly more important than surface roughness or wind speed.

3.1   Distributions of Dry Deposition Data

Dry deposition velocity is represented using linear regression with the following equation:

$$\ln(v_d) = a + b(\ln d_p) + c(\ln d_p)^2 + d(\ln d_p)^3 + e \cdot z_0 + f \cdot z_0^2 + g \cdot v \qquad (3.1)$$

where

v_d =   deposition velocity (cm/s)
d_p =   hydrodynamic particle diameter (μm)
z_0 =   surface roughness (m)
v =   wind speed (m/s)

The regression coefficients, a through g, are tabulated for each of eleven quantiles in Table 3-1. The regression was performed over the following parameter ranges: 0.1 to 10 μm for d_p, 5 to 60 cm for z_0, and 2 to 5 m/s for v. Using Equation (3.1) with parameters from Table 3-1 should provide reasonable results over these ranges. The reader is cautioned not to extrapolate too far beyond these parameter ranges. The behavior of the correlation is generally better at the 0.5 quantile and worse at the tails of the distribution. Extrapolation beyond a factor of two beyond (above or below) the parameter ranges provided above is generally discouraged. For surface roughness, even a factor of two above the indicated range (120 cm) is too much and results in non-monotonic behavior with quantile level (i.e., the 0.9 quantile may be larger than the 0.95 quantile). It is the reader's responsibility to verify that any extrapolation in the parameter ranges provides reasonable deposition velocities.

Table 3-2 shows values for deposition velocity as a function of aerosol diameter corresponding to typical choices for surface roughness (z_0 = 10 cm) and wind speed (v = 5 m/s). Figures 3-1 through 3-30 show comparisons between the original experts' data and the resampling method (RSM) results from this work. Figures 3-31 through 3-36 show comparisons of the resampled expert data with curves (labeled "Fit") constructed from Equation (3-1) using the coefficients in Table 3-1. The agreement between the expert and correlated data is generally very good.

It is not entirely obvious how to interpolate the data in Table 3-1 between quantile levels. Given the fact that many of the quantities in the table are negative, a simple linear interpolation is probably best.

Table 3-1:   Values of regression coefficients in Equation (3.1) for dry deposition velocity.

Quantile	Regression Coefficients						
	a	b	c	d	e	f	g
0.00	-6.482	1.578	-0.068	0.015	-0.056	1.235	0.071
0.01	-5.991	1.448	0.183	-0.057	-1.469	4.259	0.086
0.05	-5.504	1.121	0.284	-0.048	1.110	0.270	0.160
0.10	-5.111	1.033	0.282	-0.042	3.063	-2.703	0.155
0.25	-4.369	0.981	0.256	-0.050	3.490	-2.995	0.178
0.50	-3.112	0.992	0.190	-0.072	5.922	-6.314	0.169
0.75	-2.082	0.843	0.204	-0.045	7.768	-8.251	0.170
0.90	-1.468	0.950	0.259	-0.041	8.209	-6.243	0.218
0.95	-1.319	1.002	0.272	-0.041	16.239	-18.056	0.230
0.99	-0.601	0.934	0.266	-0.032	24.575	-31.551	0.216
1.00	0.574	0.976	0.242	-0.042	22.313	-28.886	0.191
Mean	-2.582	0.949	0.255	-0.040	15.984	-18.716	0.211
Mode	-4.838	1.003	0.333	-0.019	0.697	0.467	0.092

Table 3-2: Values of dry deposition velocity in cm/s for $z_0 = 10$ cm, $v = 5$ m/s.

Quantile	Aerosol Diameter (μm)				
	0.1	0.3	1.	3.	10
0.00	$3.4 \cdot 10^{-5}$	$2.9 \cdot 10^{-4}$	$2.2 \cdot 10^{-3}$	$1.2 \cdot 10^{-2}$	$6.9 \cdot 10^{-2}$
0.01	$6.5 \cdot 10^{-4}$	$8.7 \cdot 10^{-4}$	$3.5 \cdot 10^{-3}$	$2.0 \cdot 10^{-2}$	$1.3 \cdot 10^{-1}$
0.05	$6.2 \cdot 10^{-3}$	$4.3 \cdot 10^{-3}$	$1.0 \cdot 10^{-2}$	$4.6 \cdot 10^{-2}$	$3.4 \cdot 10^{-1}$
0.10	$1.2 \cdot 10^{-2}$	$8.1 \cdot 10^{-3}$	$1.7 \cdot 10^{-2}$	$7.2 \cdot 10^{-2}$	$5.0 \cdot 10^{-1}$
0.25	$3.2 \cdot 10^{-2}$	$2.1 \cdot 10^{-2}$	$4.3 \cdot 10^{-2}$	$1.6 \cdot 10^{-1}$	$8.6 \cdot 10^{-1}$
0.50	$1.2 \cdot 10^{-1}$	$7.9 \cdot 10^{-2}$	$1.8 \cdot 10^{-1}$	$6.0 \cdot 10^{-1}$	$2.0 \cdot 10^{0}$
0.75	$4.3 \cdot 10^{-1}$	$3.1 \cdot 10^{-1}$	$5.9 \cdot 10^{-1}$	$1.8 \cdot 10^{0}$	$7.0 \cdot 10^{0}$
0.90	$1.1 \cdot 10^{0}$	$7.3 \cdot 10^{-1}$	$1.5 \cdot 10^{0}$	$5.4 \cdot 10^{0}$	$3.1 \cdot 10^{1}$
0.95	$2.5 \cdot 10^{0}$	$1.7 \cdot 10^{0}$	$3.6 \cdot 10^{0}$	$1.4 \cdot 10^{1}$	$9.2 \cdot 10^{1}$
0.99	$9.7 \cdot 10^{0}$	$7.0 \cdot 10^{0}$	$1.4 \cdot 10^{1}$	$5.1 \cdot 10^{1}$	$3.3 \cdot 10^{2}$
1.00	$2.1 \cdot 10^{1}$	$1.5 \cdot 10^{1}$	$3.2 \cdot 10^{1}$	$1.2 \cdot 10^{2}$	$6.6 \cdot 10^{2}$
Mean	$6.3 \cdot 10^{-1}$	$4.4 \cdot 10^{-1}$	$8.9 \cdot 10^{-1}$	$3.3 \cdot 10^{0}$	$1.9 \cdot 10^{1}$
Mode	$9.9 \cdot 10^{-3}$	$6.8 \cdot 10^{-3}$	$1.4 \cdot 10^{-2}$	$5.9 \cdot 10^{-2}$	$6.3 \cdot 10^{-1}$

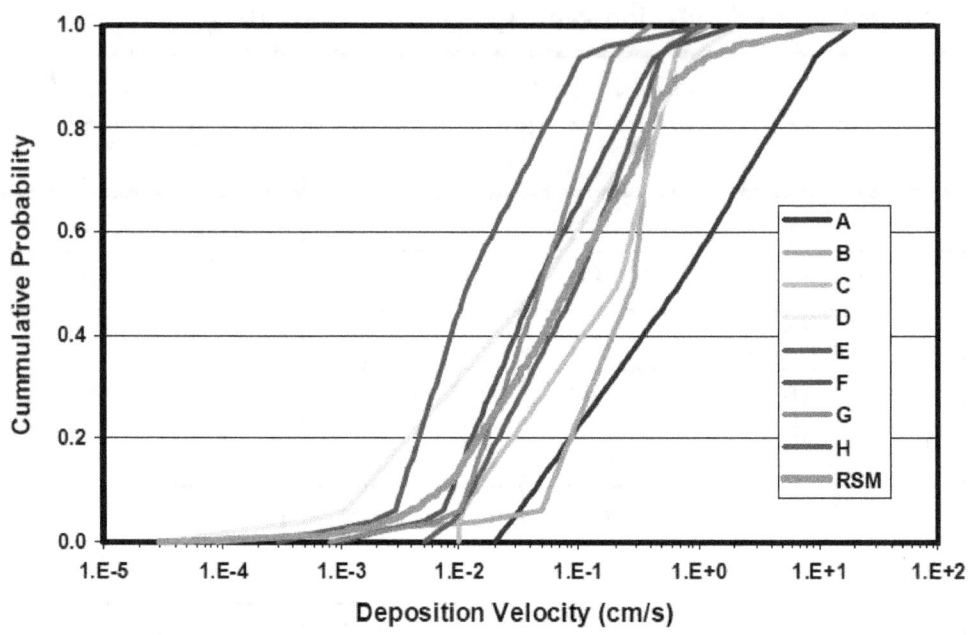

Figure 3-1. Expert and resampled data for deposition velocity corresponding to 0.1 μm particles over an urban terrain and a wind speed of 2 m/s

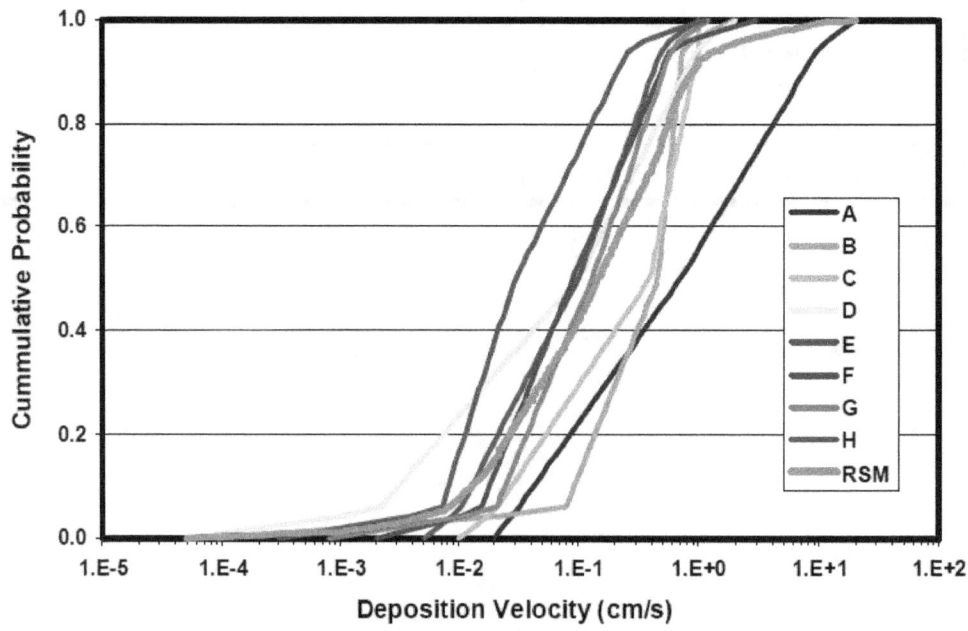

Figure 3-2. Expert and resampled data for deposition velocity corresponding to 0.1 μm particles over an urban terrain and a wind speed of 5 m/s

48

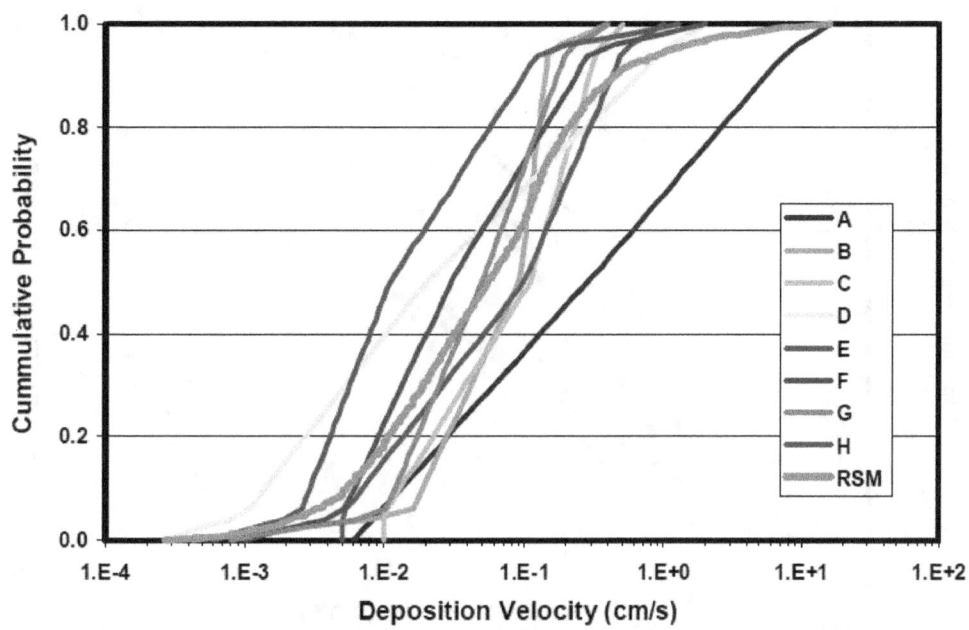

Figure 3-3. Expert and resampled data for deposition velocity corresponding to 0.3 μm particles over an urban terrain and a wind speed of 2 m/s

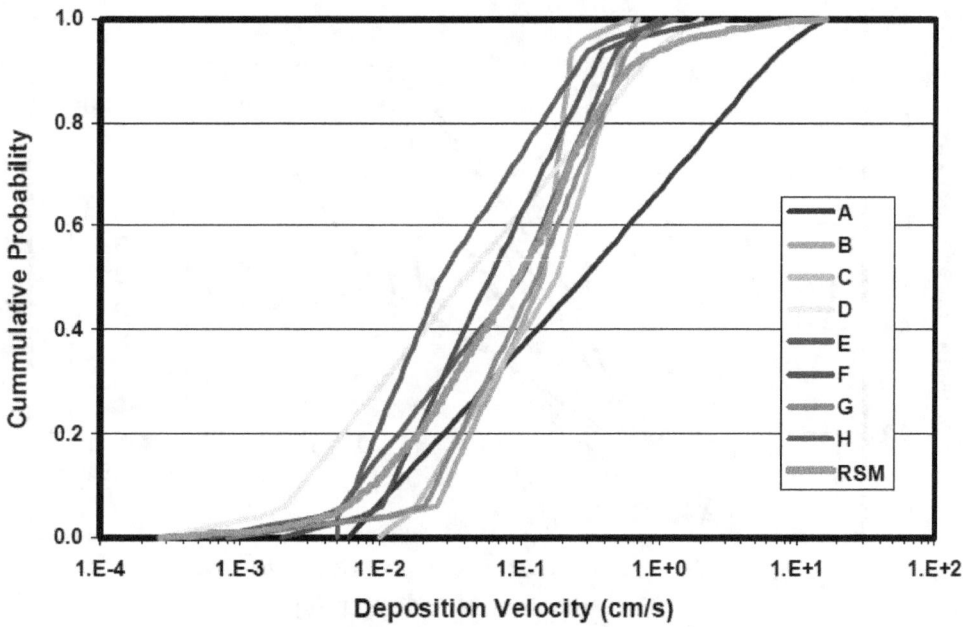

Figure 3-4. Expert and resampled data for deposition velocity corresponding to 0.3 μm particles over an urban terrain and a wind speed of 5 m/s

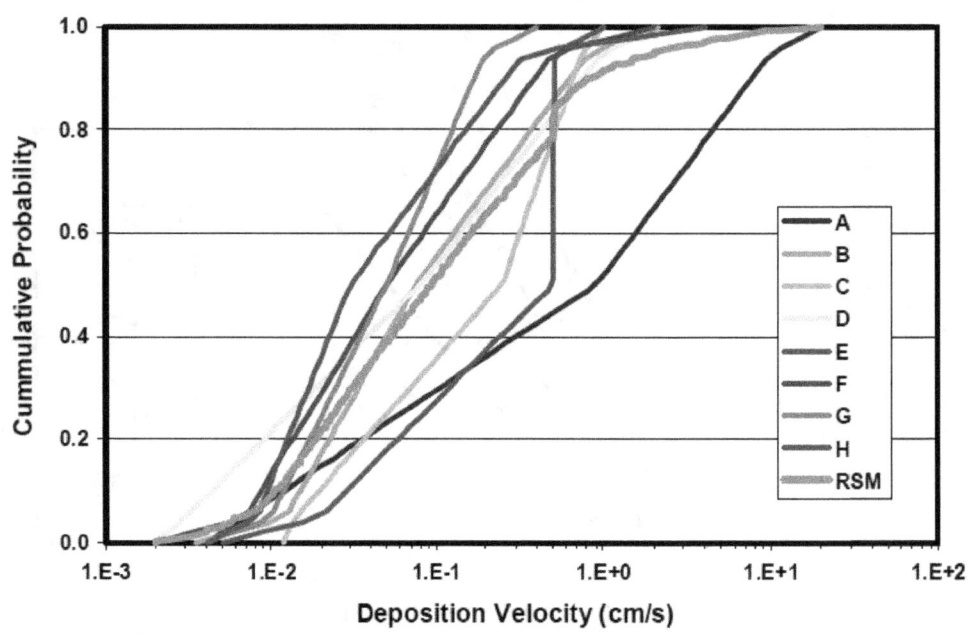

Figure 3-5. Expert and resampled data for deposition velocity corresponding to 1 μm particles over an urban terrain and a wind speed of 2 m/s

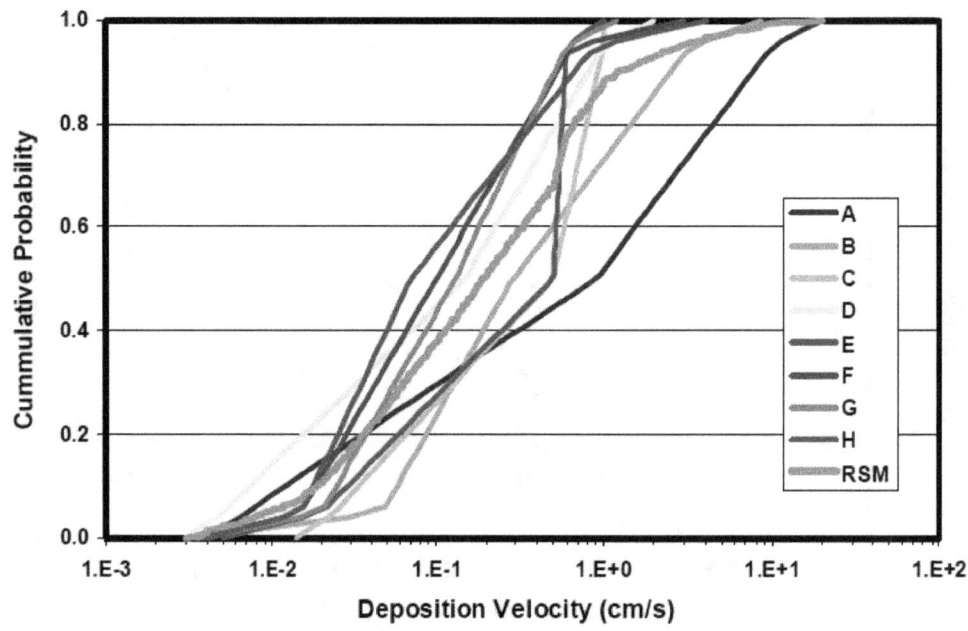

Figure 3-6. Expert and resampled data for deposition velocity corresponding to 1 μm particles over an urban terrain and a wind speed of 5 m/s

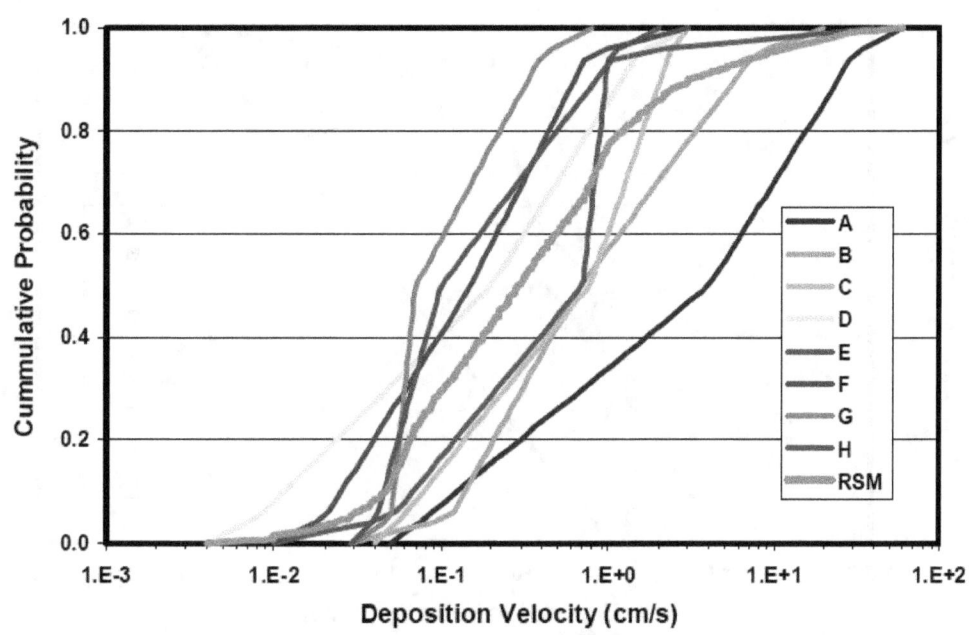

Figure 3-7. Expert and resampled data for deposition velocity corresponding to 3 μm particles over an urban terrain and a wind speed of 2 m/s

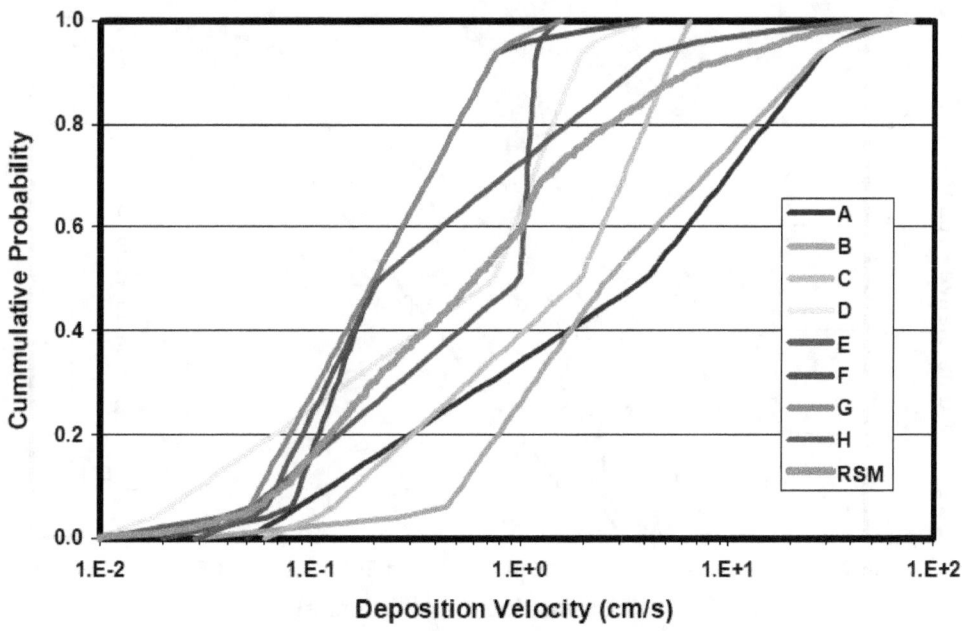

Figure 3-8. Expert and resampled data for deposition velocity corresponding to 3 μm particles over an urban terrain and a wind speed of 5 m/s

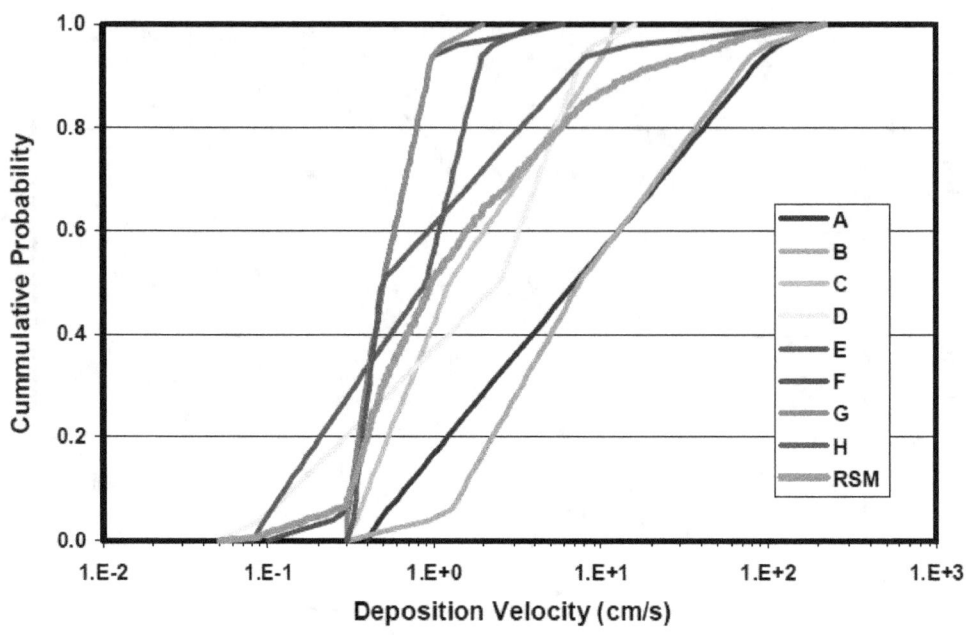

Figure 3-9. **Expert and resampled data for deposition velocity corresponding to 10 μm particles over an urban terrain and a wind speed of 2 m/s**

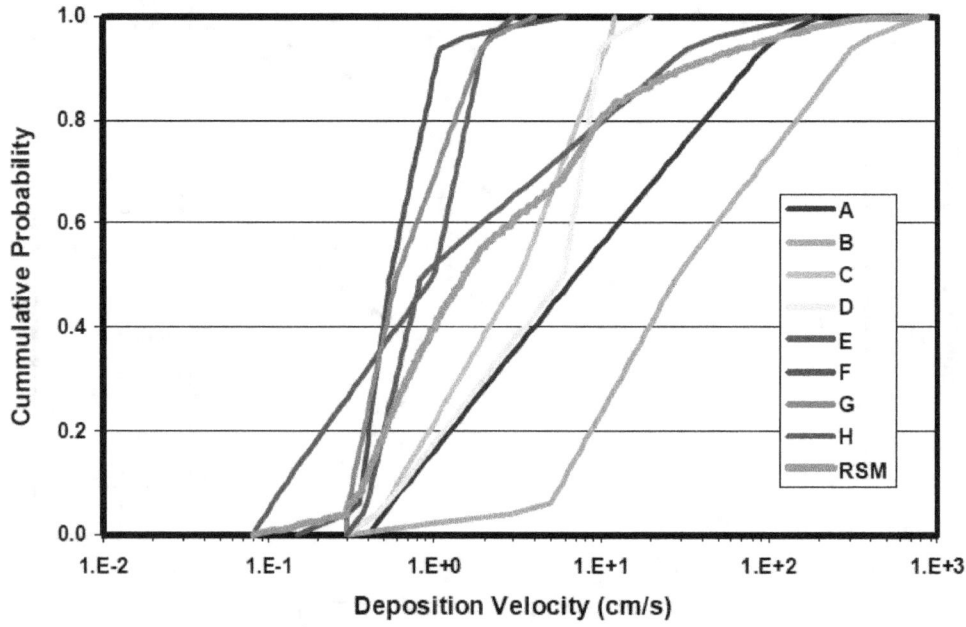

Figure 3-10. **Expert and resampled data for deposition velocity corresponding to 10 μm particles over an urban terrain and a wind speed of 5 m/s**

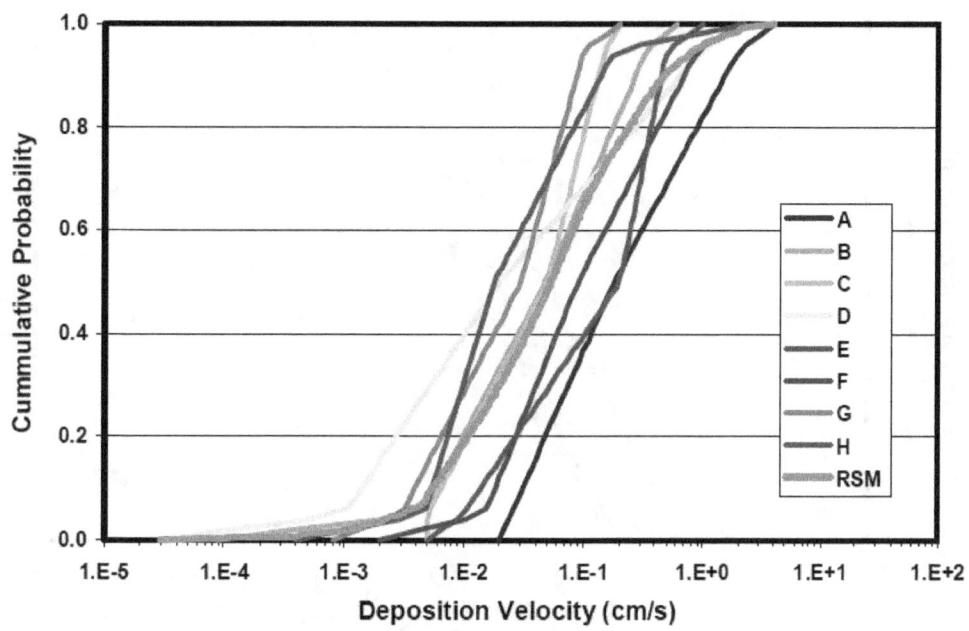

Figure 3-11. Expert and resampled data for deposition velocity corresponding to 0.1 μm particles over a meadow and a wind speed of 2 m/s

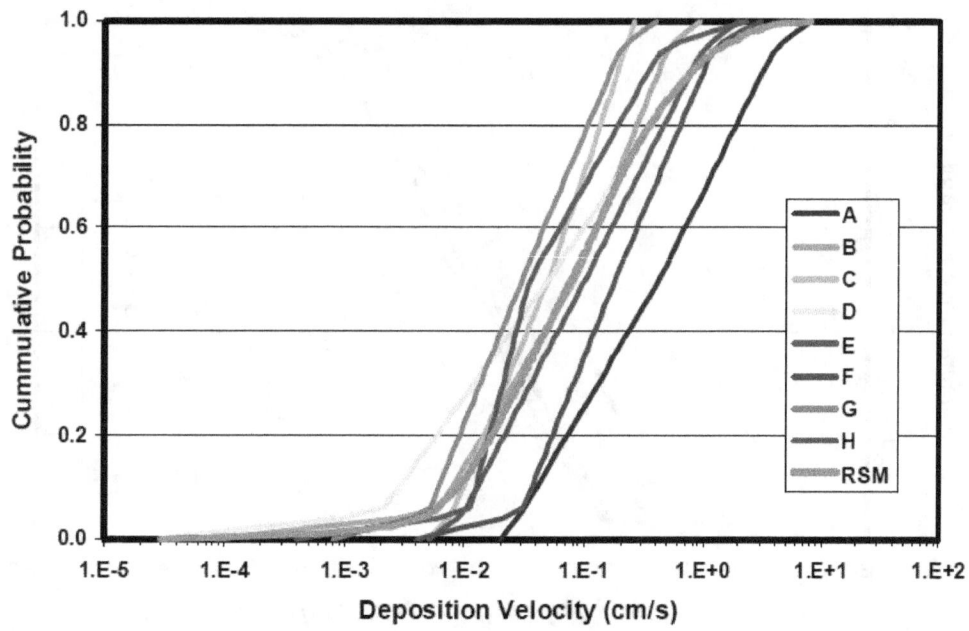

Figure 3-12. Expert and resampled data for deposition velocity corresponding to 0.1 μm particles over a meadow and a wind speed of 5 m/s

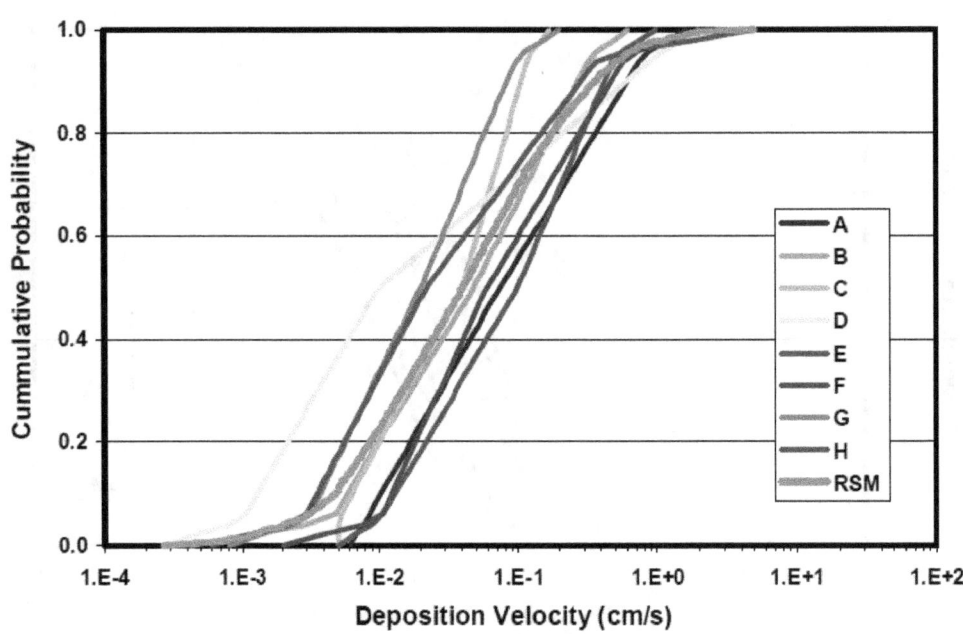

Figure 3-13. **Expert and resampled data for deposition velocity corresponding to 0.3 μm particles over a meadow and a wind speed of 2 m/s**

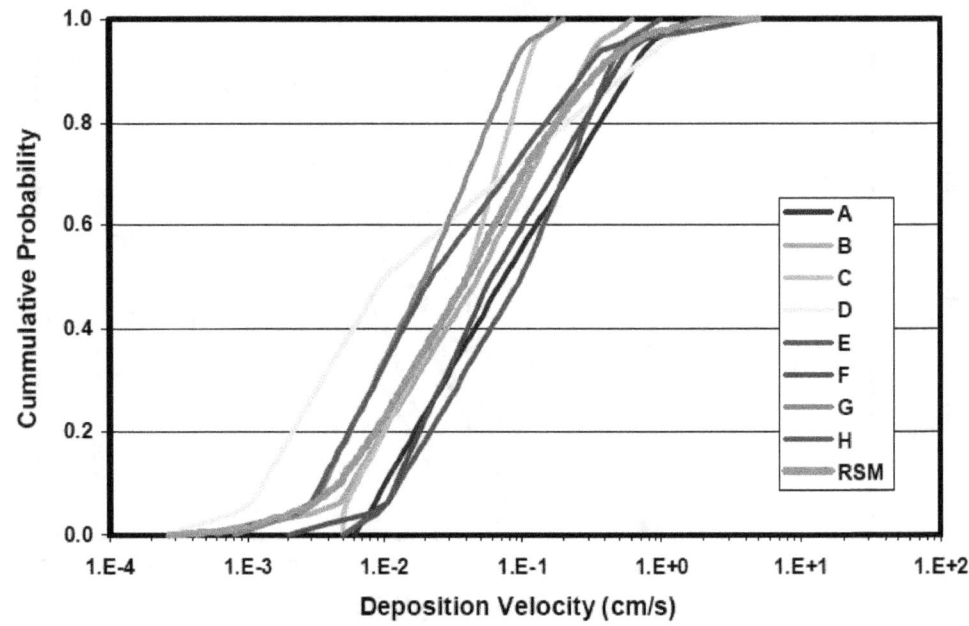

Figure 3-14. **Expert and resampled data for deposition velocity corresponding to 0.3 μm particles over a meadow and a wind speed of 5 m/s**

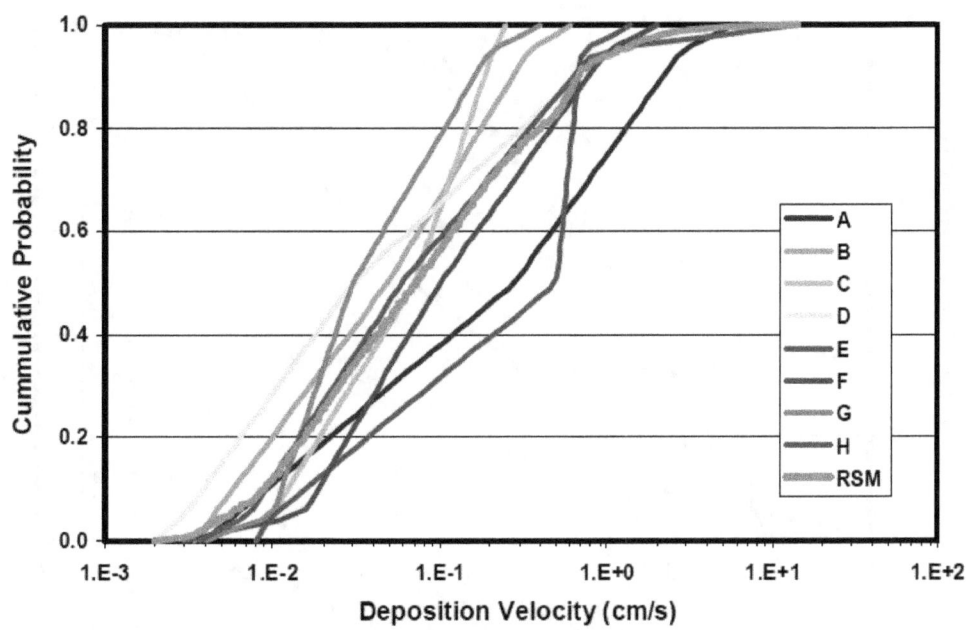

Figure 3-15. Expert and resampled data for deposition velocity corresponding to 1 μm particles over a meadow and a wind speed of 2 m/s

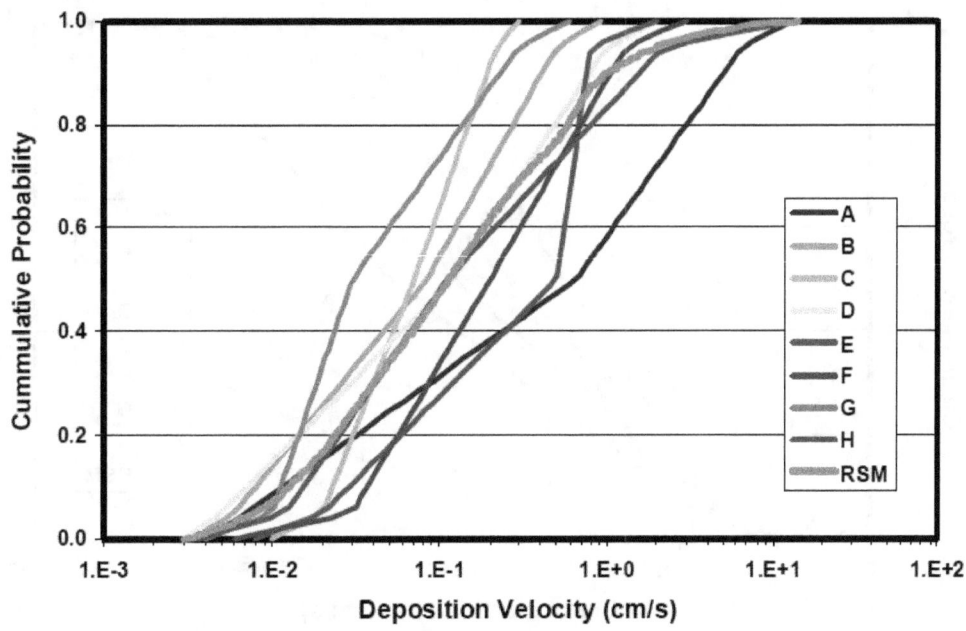

Figure 3-16. Expert and resampled data for deposition velocity corresponding to 1 μm particles over a meadow and a wind speed of 5 m/s

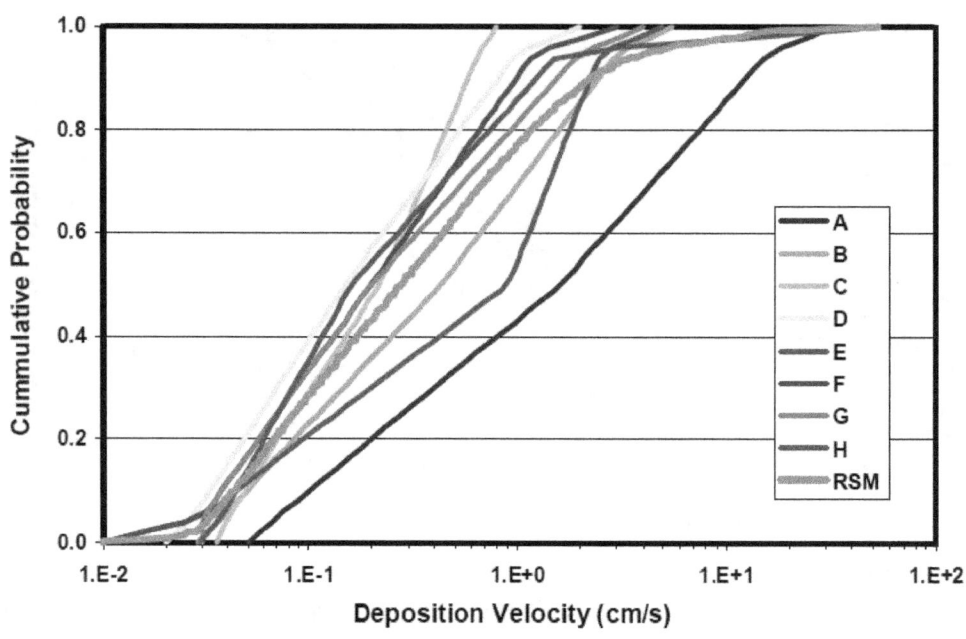

Figure 3-17. Expert and resampled data for deposition velocity corresponding to 3 μm particles over a meadow and a wind speed of 2 m/s

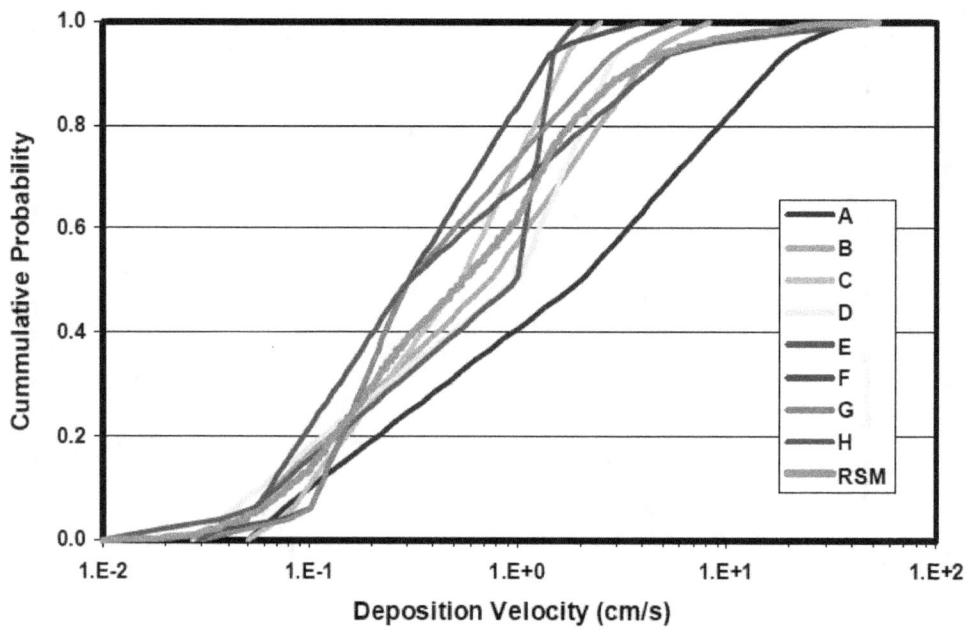

Figure 3-18. Expert and resampled data for deposition velocity corresponding to 3 μm particles over a meadow and a wind speed of 5 m/s

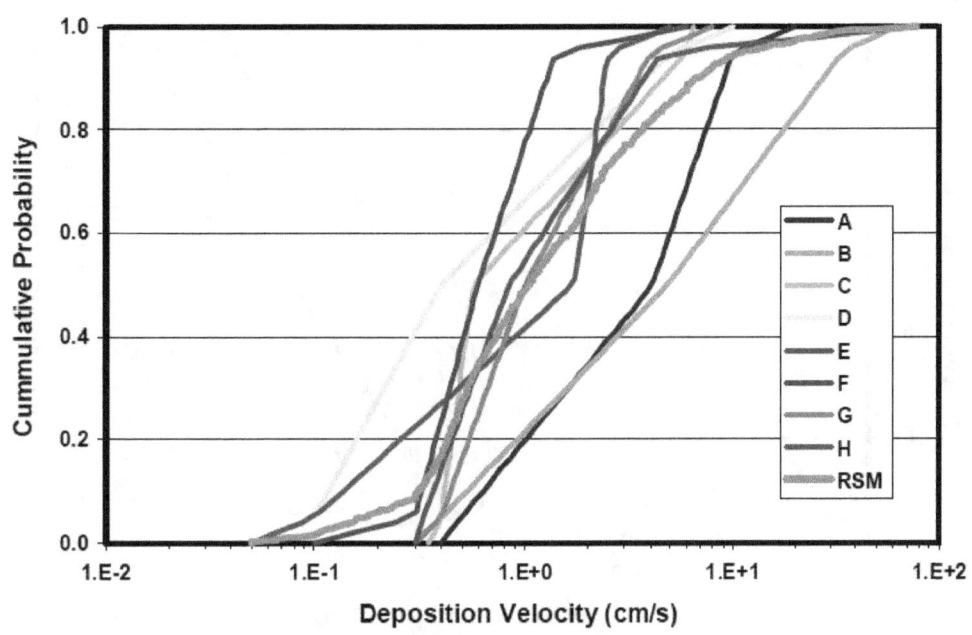

Figure 3-19. Expert and resampled data for deposition velocity corresponding to 10 μm particles over a meadow and a wind speed of 2 m/s

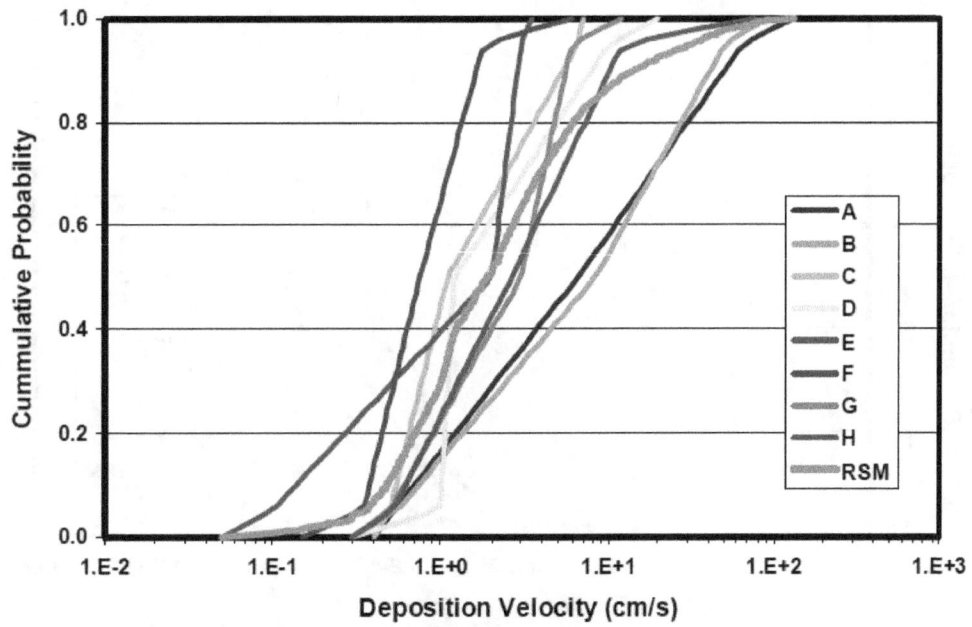

Figure 3-20. Expert and resampled data for deposition velocity corresponding to 10 μm particles over a meadow and a wind speed of 5 m/s

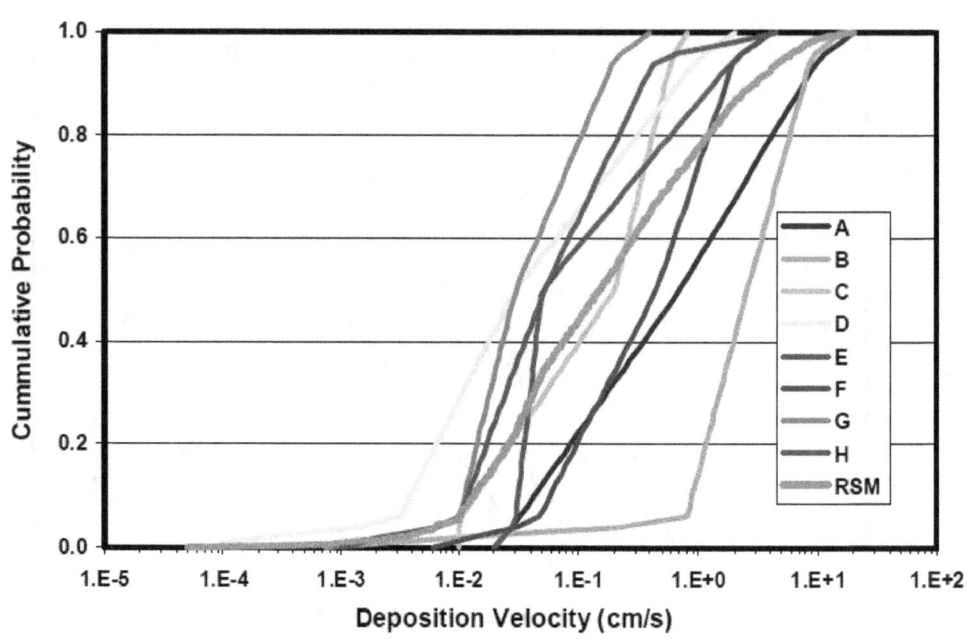

Figure 3-21. **Expert and resampled data for deposition velocity corresponding to 0.1 μm particles over a forrest and a wind speed of 2 m/s**

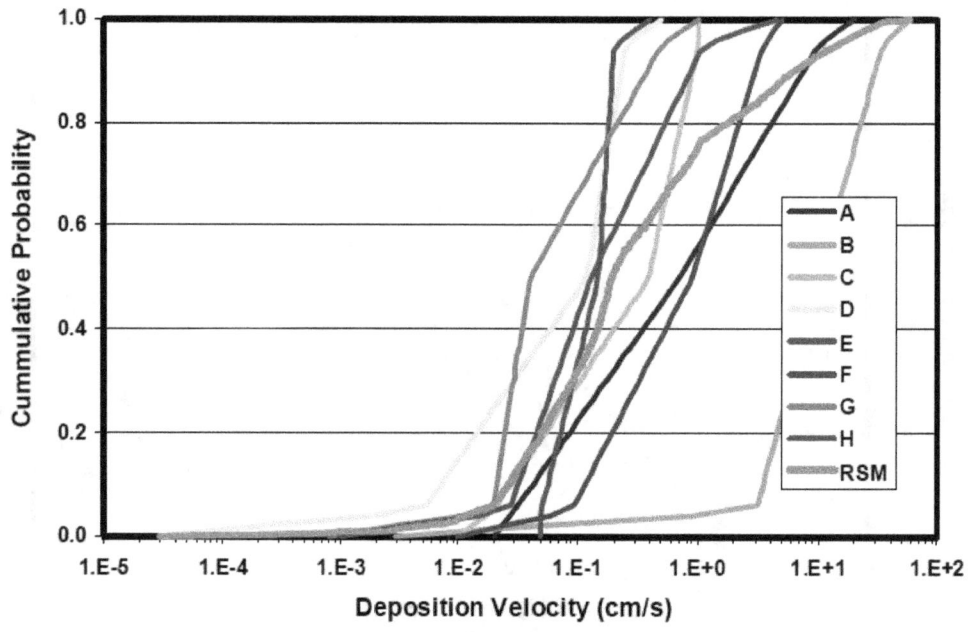

Figure 3-22. **Expert and resampled data for deposition velocity corresponding to 0.1 μm particles over a forrest and a wind speed of 5 m/s**

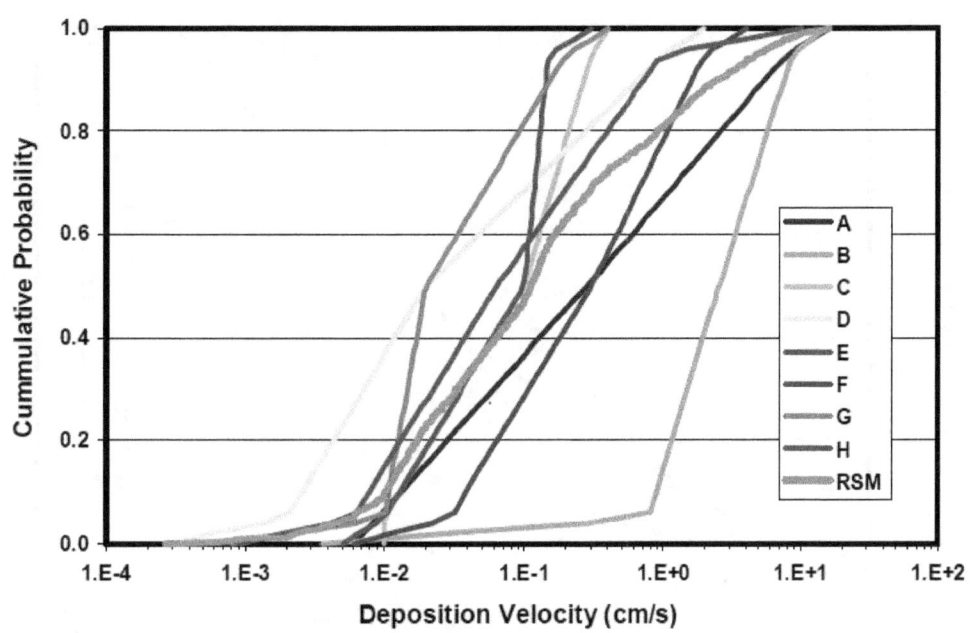

Figure 3-23. Expert and resampled data for deposition velocity corresponding to 0.3 μm particles over a forrest and a wind speed of 2 m/s

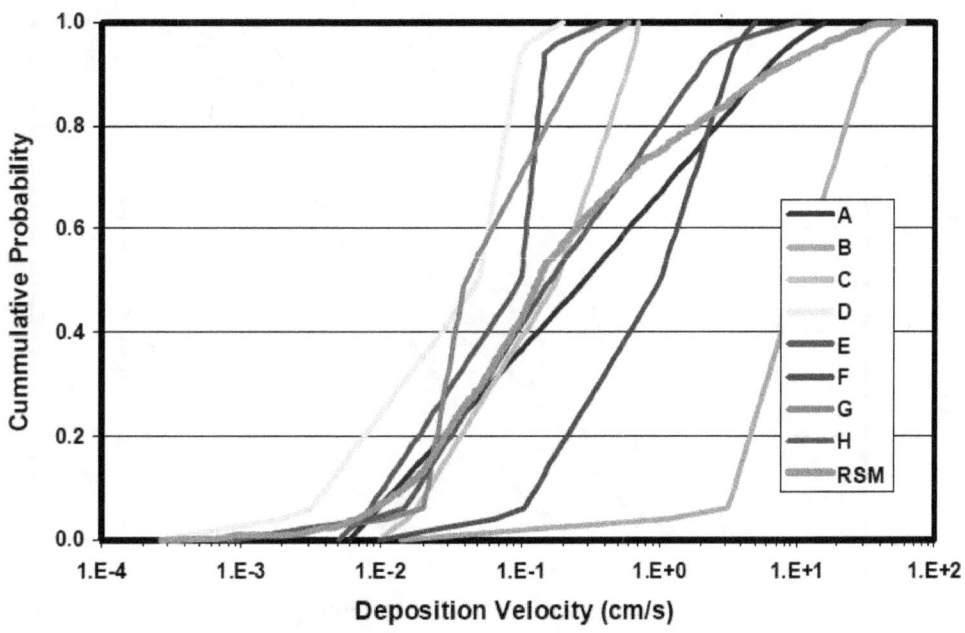

Figure 3-24. Expert and resampled data for deposition velocity corresponding to 0.3 μm particles over a forrest and a wind speed of 5 m/s

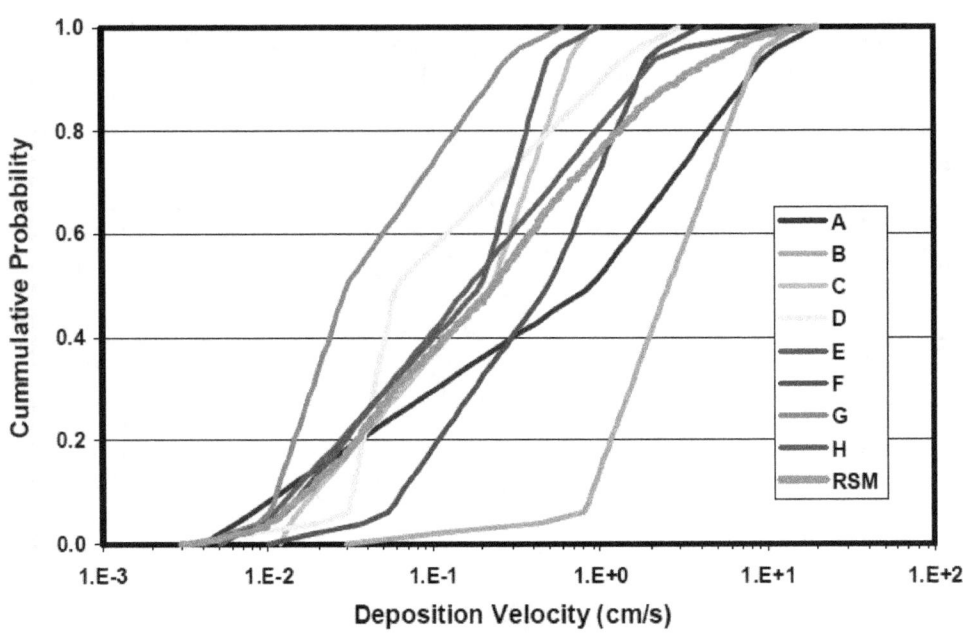

Figure 3-25. Expert and resampled data for deposition velocity corresponding to 1 μm particles over a forrest and a wind speed of 2 m/s

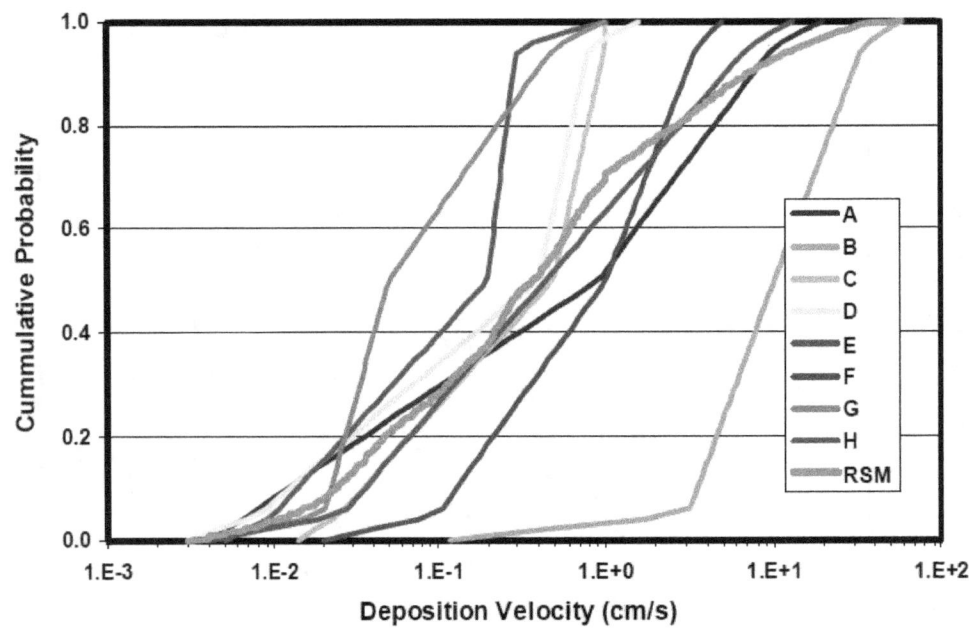

Figure 3-26. Expert and resampled data for deposition velocity corresponding to 1 μm particles over a forrest and a wind speed of 5 m/s

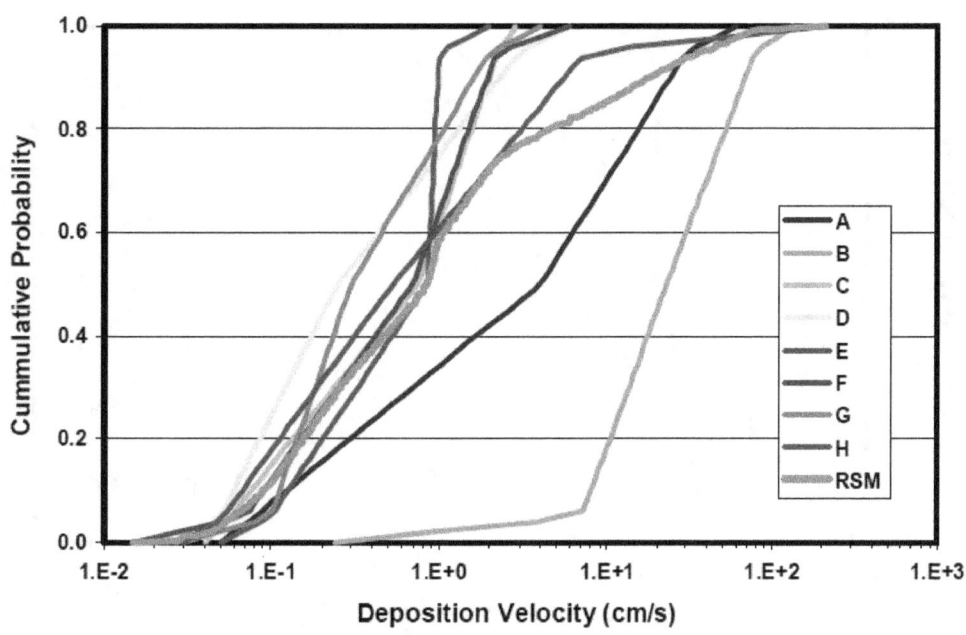

Figure 3-27. Expert and resampled data for deposition velocity corresponding to 3 μm particles over a forrest and a wind speed of 2 m/s

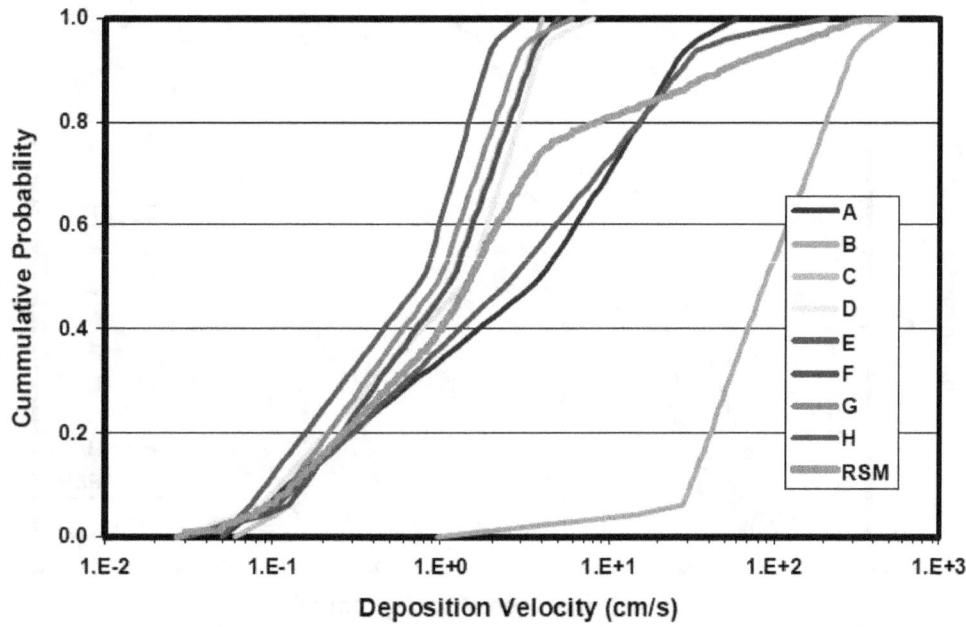

Figure 3-28. Expert and resampled data for deposition velocity corresponding to 3 μm particles over a forrest and a wind speed of 5 m/s

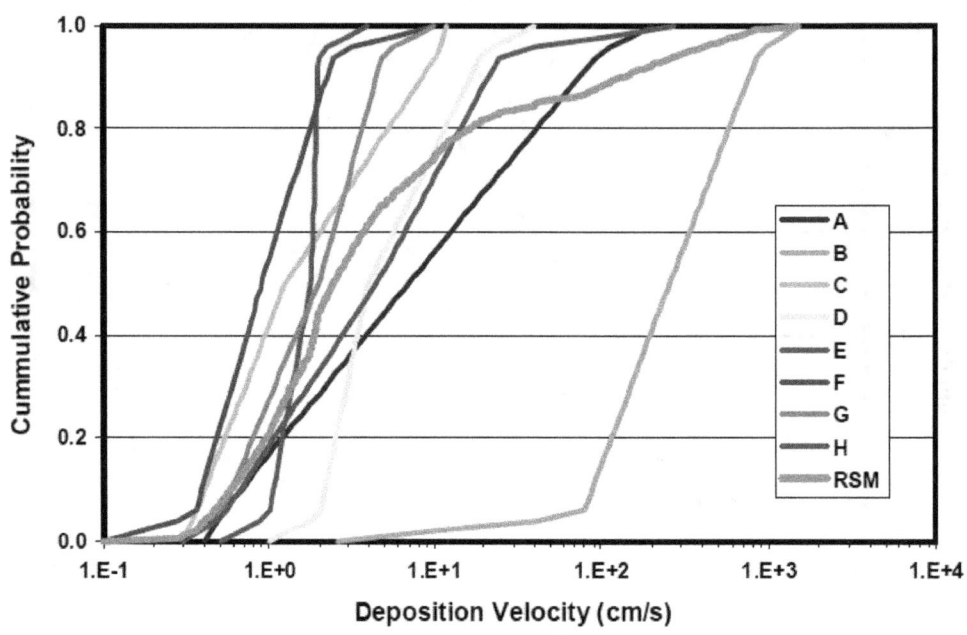

Figure 3-29. **Expert and resampled data for deposition velocity corresponding to 10 μm particles over a forrest and a wind speed of 2 m/s**

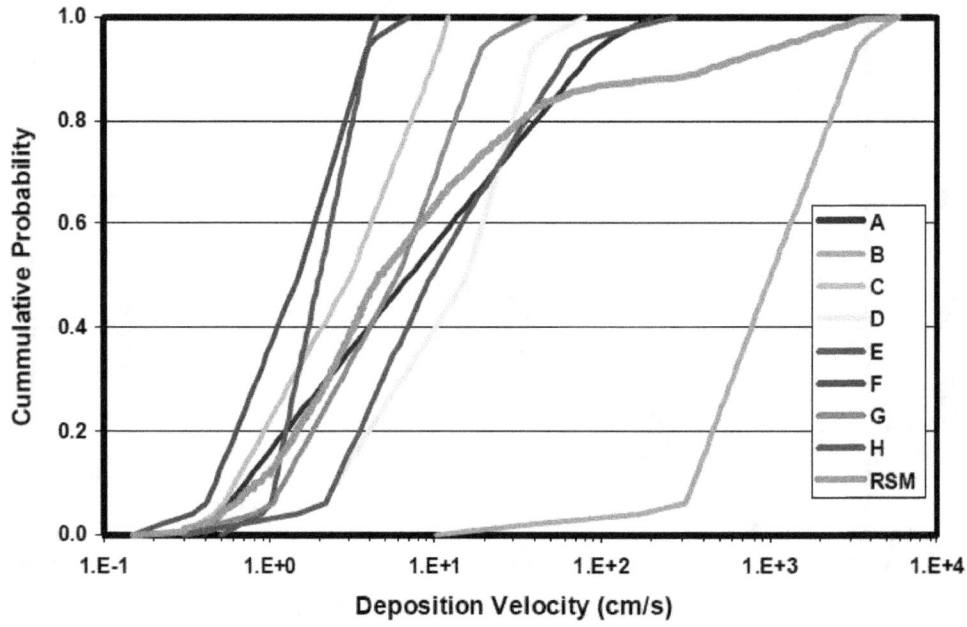

Figure 3-30. **Expert and resampled data for deposition velocity corresponding to 10 μm particles over a forrest and a wind speed of 5 m/s**

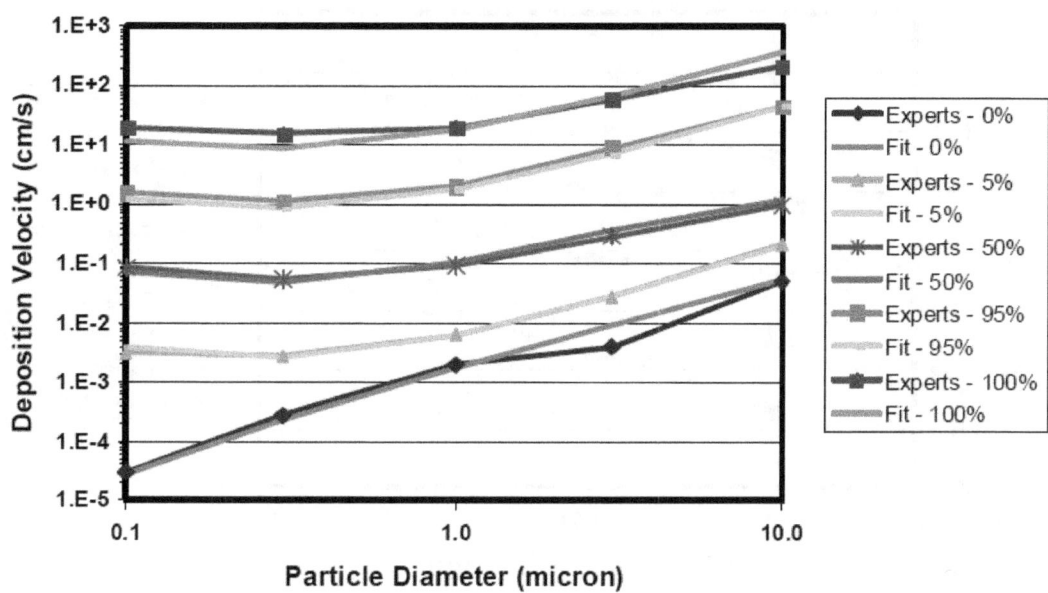

Figure 3-31. Comparison of resampled expert data for dry deposition velocity over urban terrain and wind speed of 2 m/s with the calculated values using the parameters in Table 3-1

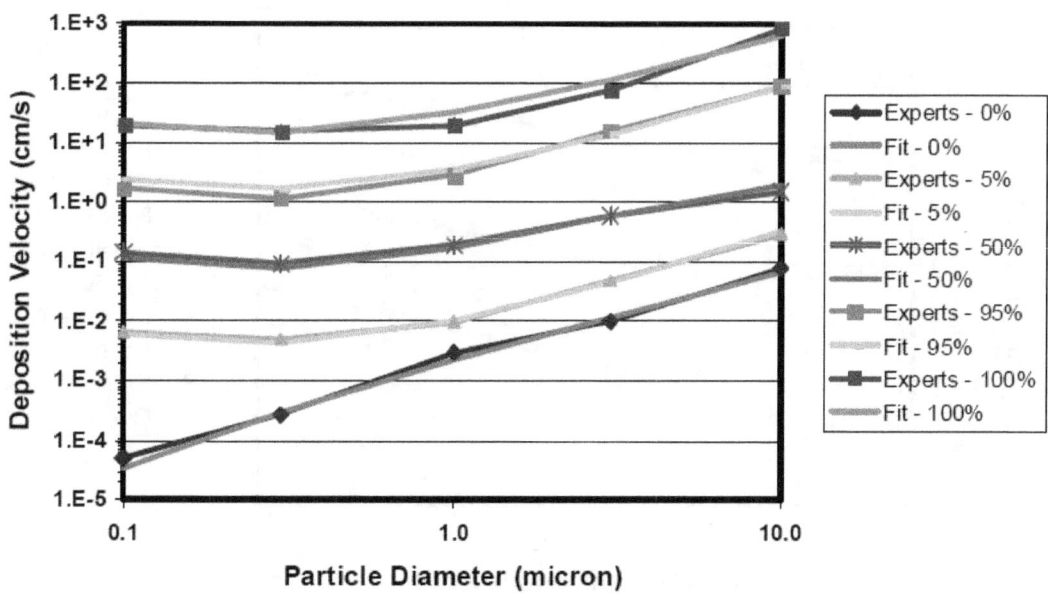

Figure 3-32. Comparison of resampled expert data for dry deposition velocity over urban terrain and wind speed of 5 m/s with the calculated values using the parameters in Table 3-1

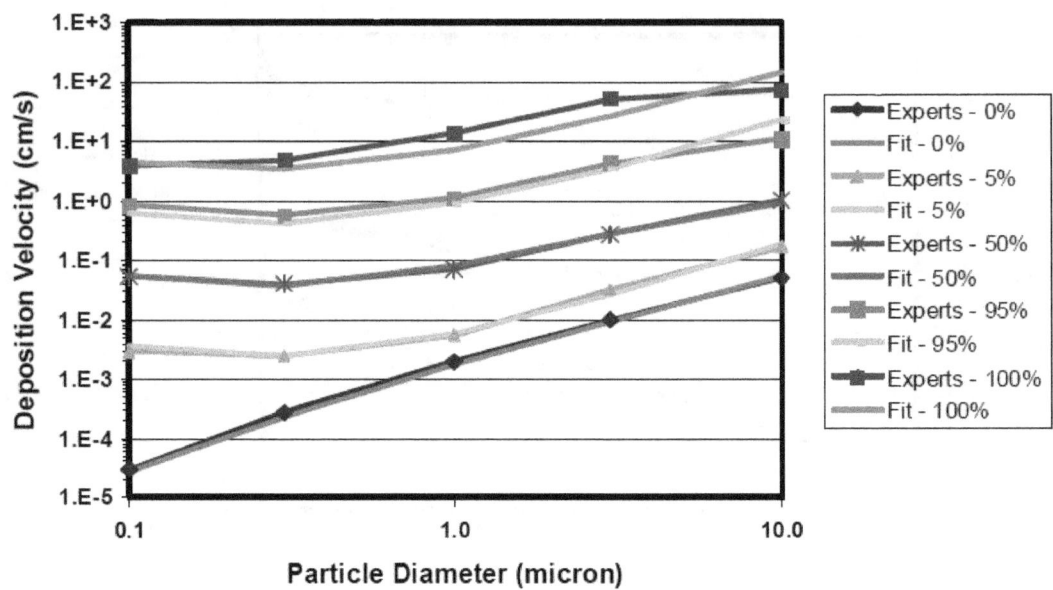

Figure 3-33. Comparison of resampled expert data for dry deposition velocity over meadow and wind speed of 2 m/s with the calculated values using the parameters in Table 3-1

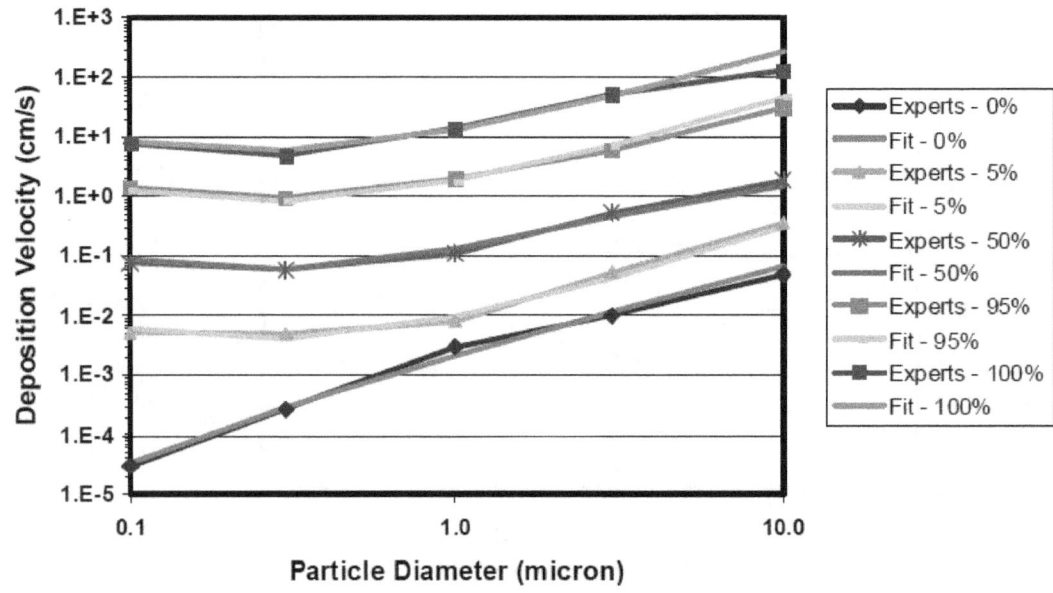

Figure 3-34. Comparison of resampled expert data for dry deposition velocity over meadow and wind speed of 5 m/s with the calculated values using the parameters in Table 3-1

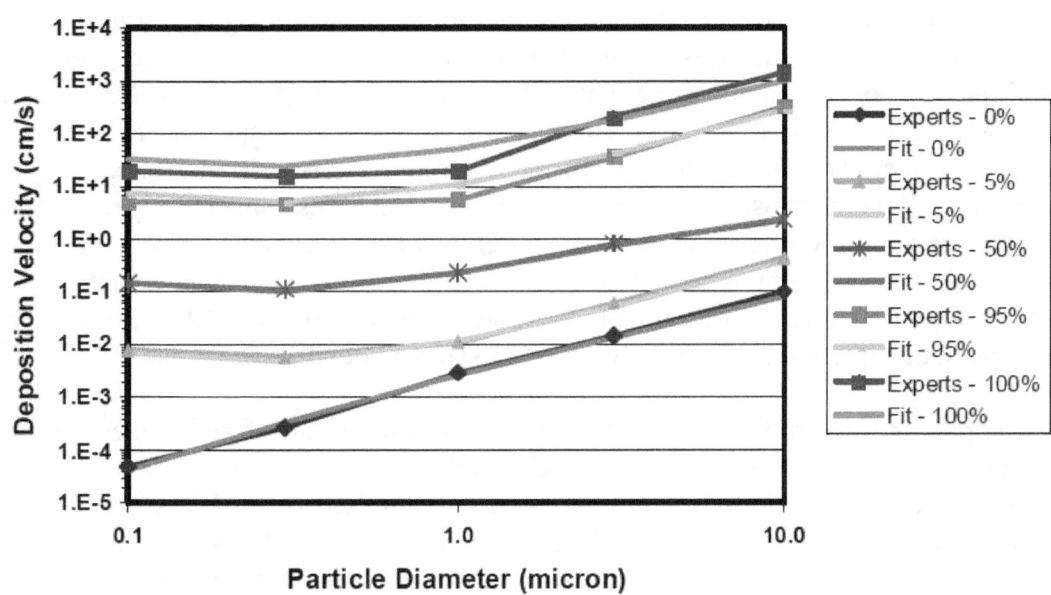

Figure 3-35. Comparison of resampled expert data for dry deposition velocity over forrest and wind speed of 2 m/s with the calculated values using the parameters in Table 3-1

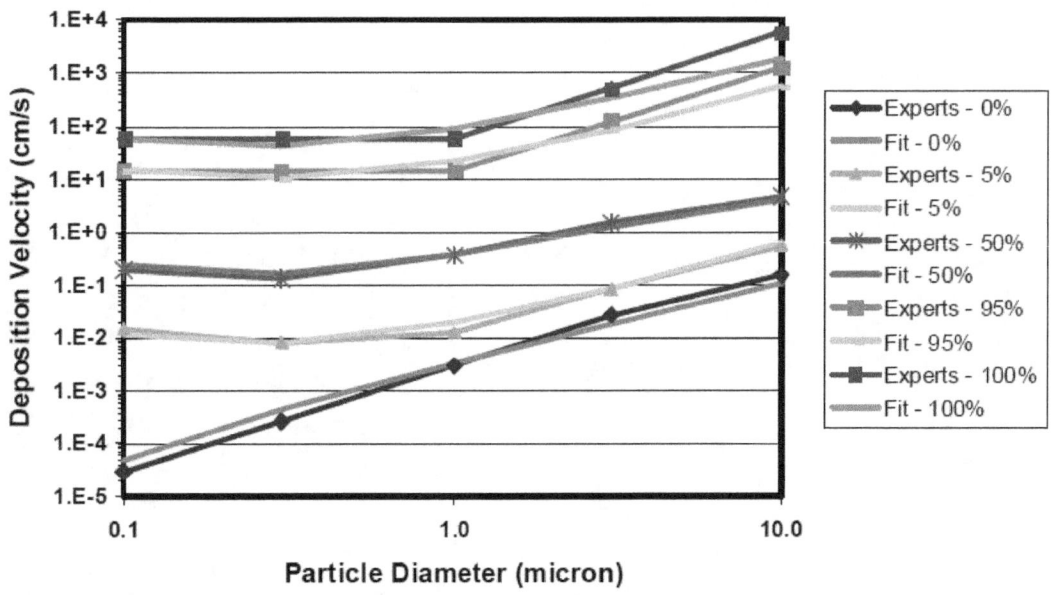

Figure 3-36. Comparison of resampled expert data for dry deposition velocity over forrest and wind speed of 5 m/s with the calculated values using the parameters in Table 3-1

3.2   Correlations Between Dry Deposition Data

Sampling for dry deposition velocities could be performed by sampling any or all of the parameters in Equation 3.1; this includes hydrodynamic particle diameter, surface roughness, wind speed, and quantile level. Quantile level enters Equation 3.1 implicitly through the values of the linear coefficients, a, b c, d, e, f, and g. However, it is reasonable that, for each realization, the same value of surface roughness, wind speed, and quantile level be selected for each of the hydrodynamic particle diameters used in the calculation. This ensures that the expected relationship between deposition velocity and particle diameter is preserved within each realization. This approach amounts to assuming a perfect correlation between the deposition velocities for a set of aerosol diameters. In practice, a correlation coefficient between 0.9 and 1.0 is probably sufficient to ensure a reasonable relationship between aerosol size and deposition velocity.

A simpler approach for sampling deposition velocities is to assume fixed values for each of the aerosol diameters, surface roughness, and wind speed. In this case, the aerosol diameters would be fixed and a set of deposition velocities would be sampled for each aerosol diameter. This approach is sufficient because quantile level has a much stronger influence on deposition velocity than any of the other variables in Equation 3.1. Again, it is reasonable to assume a correlation coefficient between 0.9 to 1.0 between each of the distributions representing the set of aerosol diameters to ensure that a correct relationship between aerosol size and deposition velocity exists within each of the realizations.

4.0 WET DEPOSITION

4.1   Distributions of Wet-Deposition Data

Wet deposition is modeled in MACCS2 using the following formula:

$$R = C_1 I^{C_2} \qquad\qquad (4.1)$$

where
R = the fractional rate at which aerosols are removed from the plume (s^{-1})
C_1 = an empirical, linear coefficient (s^{-1})
I = rain intensity (mm/hr)
C_2 = an empirical exponent (dimensionless)

The first attempt to analyze the expert data allowed both C_1 and C_2 to be functions of quantile. The values of C_1 turned out to increase monotonically with quantile; the values of C_2 did not. Because C_1 and C_2 are correlated, the results could not be implemented with latin hypercube sampling (LHS). As a result, a single value of C_2 was obtained, independent of quantile. The results are shown in Table 4-1.

The more general result obtained from the experts' data for C_1 is a correlation of the following form:

$$C_1 = a + b \cdot \ln d_p + c \cdot (\ln d_p)^2 + d \cdot (\ln d_p)^3 \qquad\qquad (4.2)$$

The regression coefficients in Equation (4.2) are displayed in Table 4-2 for each quantile level. The correlation is over the particle diameter range of 0.1 to 10 μm that was used in the expert elicitation process. The values of C_1 should not be extrapolated too far below or above this range. Even a factor of two above the indicated range in particle diameter (20 μm) is too much and produces non-physical results. It is the reader's responsibility to verify that any extrapolation in the parameter range provides reasonable wet deposition removal rates.

Plots showing comparisons between the expert data and the resampled data are shown in Figures 4-1 through 4-20. Comparisons between the correlations and the resampled data are provided in Figures 4-21 through 4-25. Curves labeled "Fit" on Figures 4-21 through 4-25 were generated using the parameter values in Table 4-2.

It is not entirely obvious how to interpolate the data in Tables 4-1 and 4-2 between quantile levels. Given the fact that many of the quantities in the tables are negative, a simple linear interpolation is probably best.

Table 4-1: **Values for the wet deposition coefficients in Equation (4.1) for 1 μm particles.**

Quantile	Wet Deposition Parameters	
	C_1 (s^{-1})	C_2 (dimensionless)
0.00	$2.73 \cdot 10^{-8}$	0.664
0.01	$2.92 \cdot 10^{-7}$	0.664
0.05	$9.13 \cdot 10^{-7}$	0.664
0.10	$1.73 \cdot 10^{-6}$	0.664
0.25	$5.36 \cdot 10^{-6}$	0.664
0.50	$1.89 \cdot 10^{-5}$	0.664
0.75	$9.84 \cdot 10^{-5}$	0.664
0.90	$2.59 \cdot 10^{-4}$	0.664
0.95	$5.79 \cdot 10^{-4}$	0.664
0.99	$3.78 \cdot 10^{-3}$	0.664
1.00	$1.14 \cdot 10^{-2}$	0.664
Mean	$2.21 \cdot 10^{-5}$	0.664
Mode	$4.49 \cdot 10^{-5}$	0.664

Table 4-2: Values of regression coefficients in Equation (4.2) for the wet deposition coefficient C_1.

Quantile	Regression Coefficients			
	a	b	c	d
0.00	-17.415	1.948	-0.306	-0.159
0.01	-15.045	2.701	-0.124	-0.270
0.05	-13.907	2.740	-0.034	-0.288
0.10	-13.269	2.812	-0.009	-0.307
0.25	-12.136	2.529	0.075	-0.296
0.50	-10.875	1.600	0.122	-0.145
0.75	-9.227	1.242	0.092	-0.126
0.90	-8.259	0.877	0.082	-0.076
0.95	-7.454	1.305	0.034	-0.157
0.99	-5.579	1.536	-0.058	-0.217
1.00	-4.472	0.297	-0.031	-0.043
Mean	-10.721	1.820	0.066	0.271
Mode	-10.012	0.995	0.163	-0.051

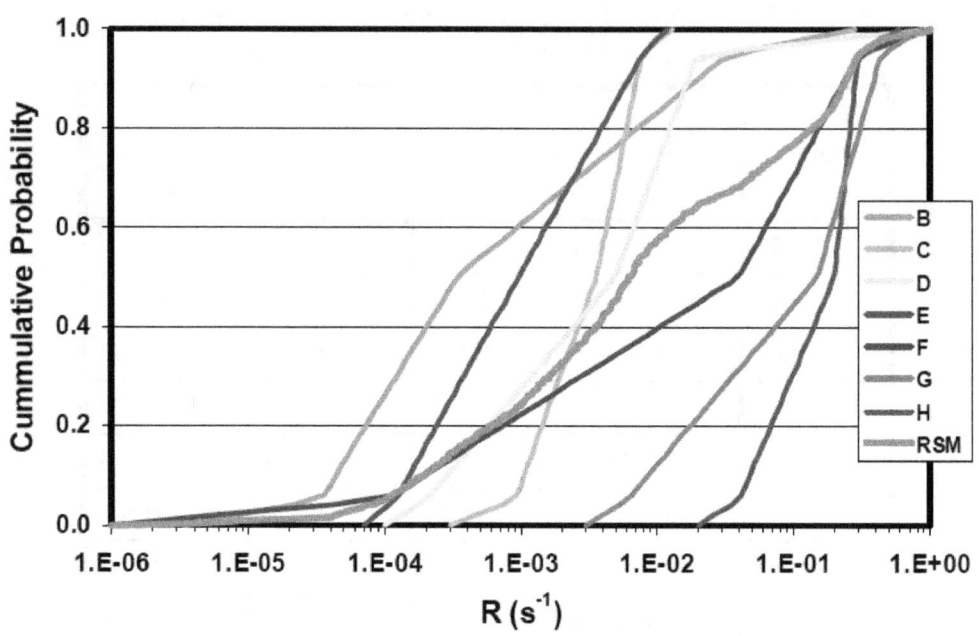

Figure 4-1.   Expert and resampled data for wet deposition rate (R in units of s^{-1}) corresponding to 0.1 μm particles and a rainfall intensity of 0.3 mm/hr

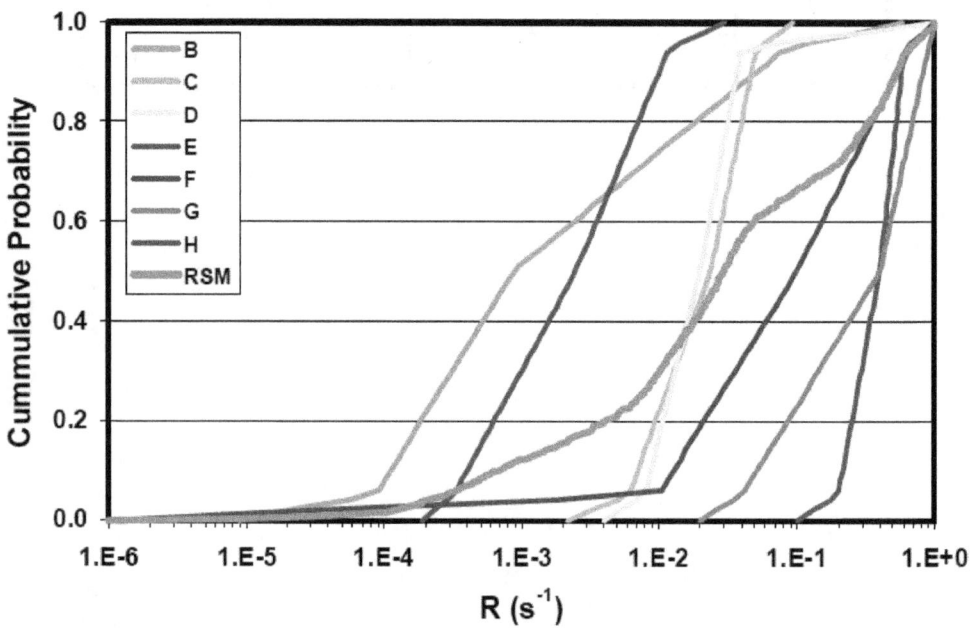

Figure 4-2.   Expert and resampled data for wet deposition rate (R in units of s^{-1}) corresponding to 0.1 μm particles and a rainfall intensity of 2 mm/hr

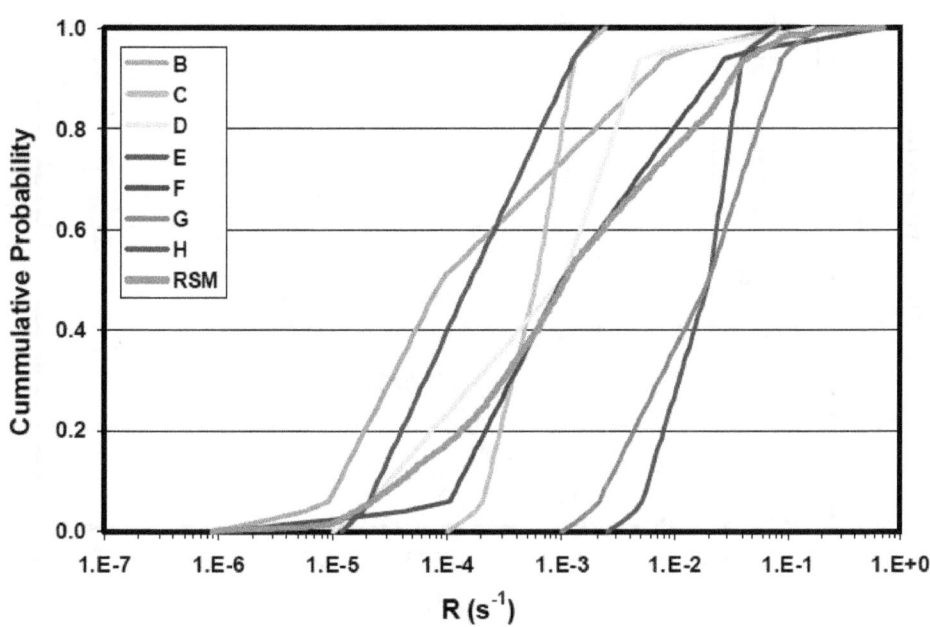

Figure 4-3. Expert and resampled data for wet deposition rate (R in units of s⁻¹) corresponding to 0.1 μm particles and a rainfall intensity of 0.05 mm/10 min

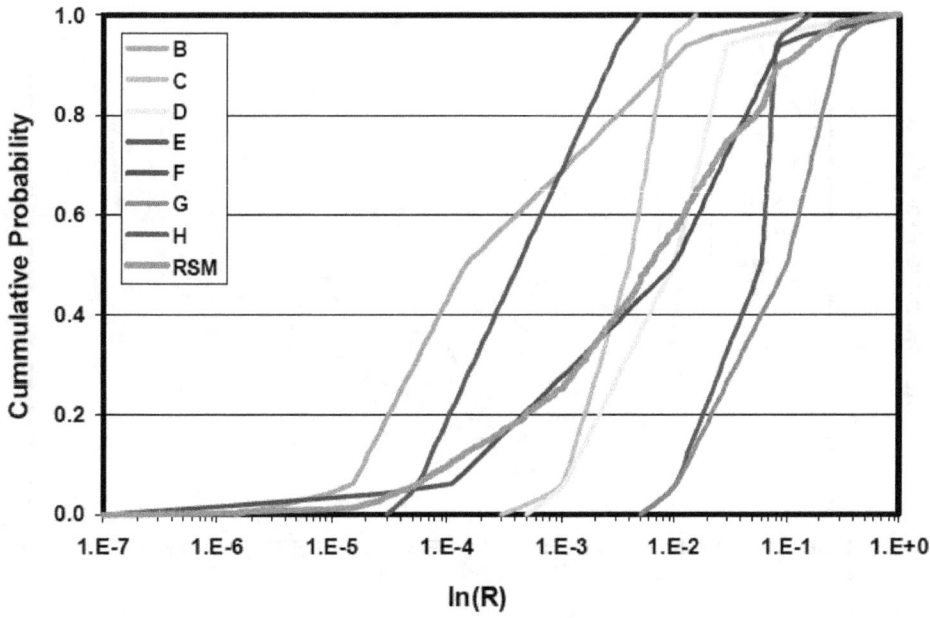

Figure 4-4. Expert and resampled data for wet deposition rate (R in units of s⁻¹) corresponding to 0.1 μm particles and a rainfall intensity of 0.33 mm/10 min

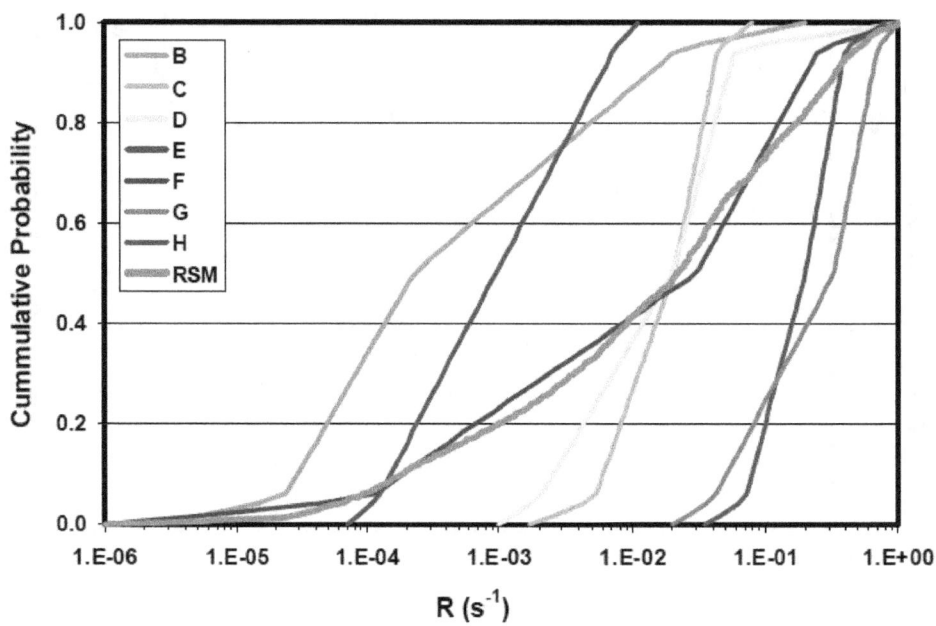

Figure 4-5. Expert and resampled data for wet deposition rate (R in units of s⁻¹) corresponding to 0.1 μm particles and a rainfall intensity of 1.67 mm/10 min

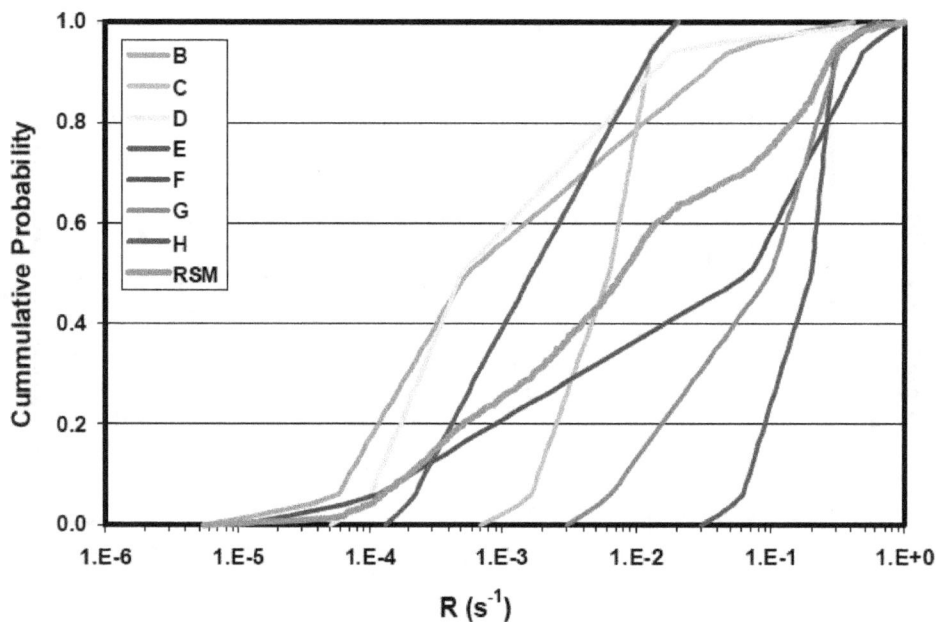

Figure 4-6. Expert and resampled data for wet deposition rate (R in units of s⁻¹) corresponding to 0.3 μm particles and a rainfall intensity of 0.3 mm/hr

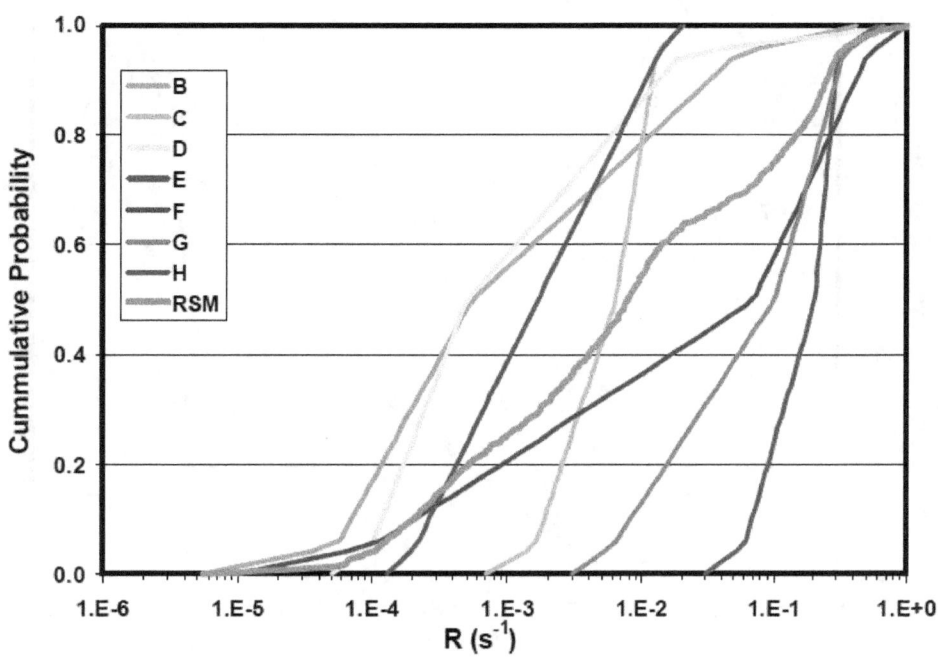

Figure 4-7. Expert and resampled data for wet deposition rate (R in units of s^{-1}) corresponding to 0.3 μm particles and a rainfall intensity of 2 mm/hr

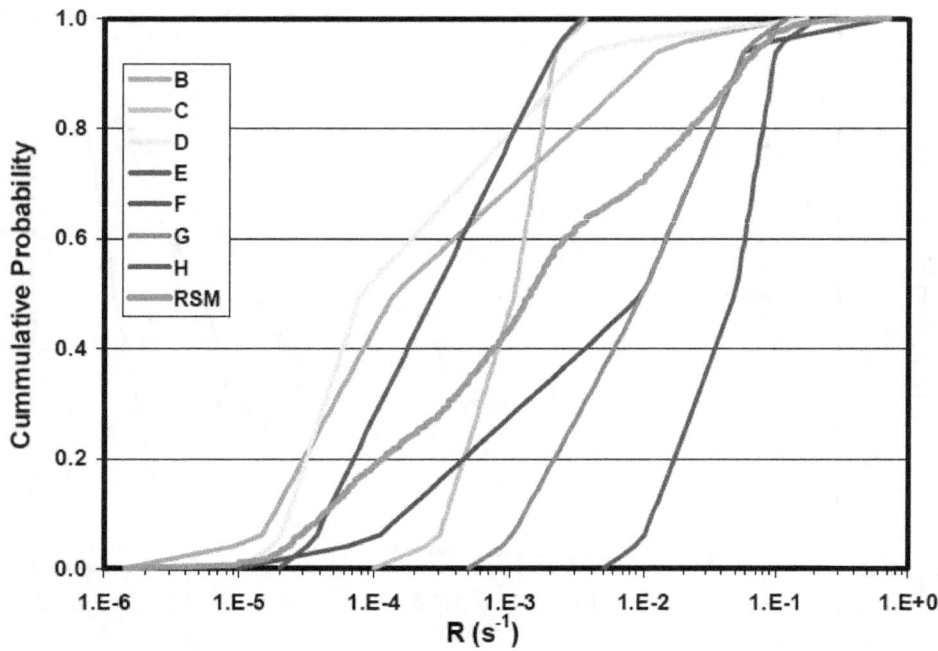

Figure 4-8. Expert and resampled data for wet deposition rate (R in units of s^{-1}) corresponding to 0.3 μm particles and a rainfall intensity of 0.05 mm/10 min

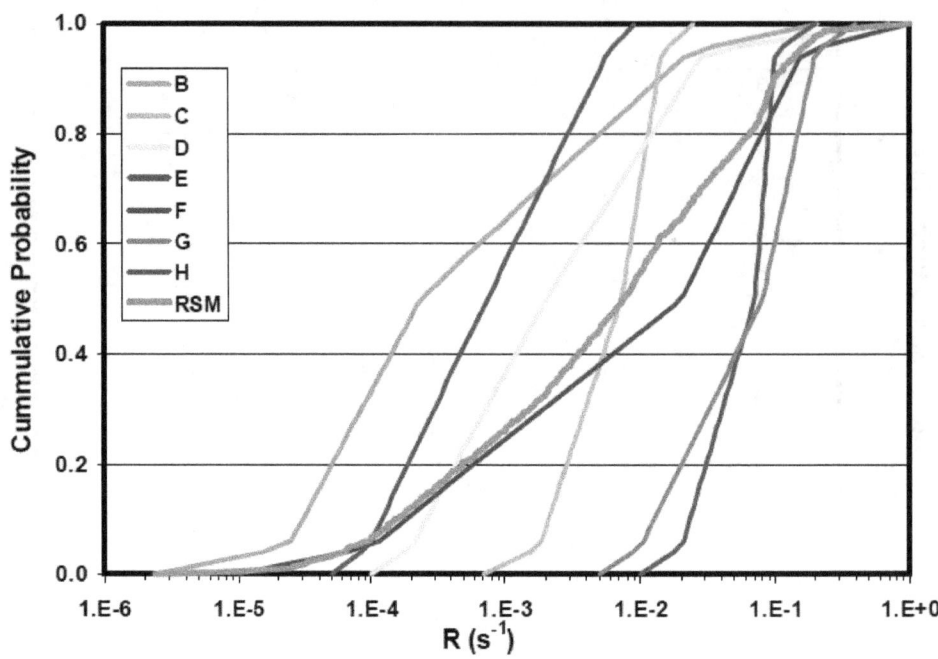

Figure 4-9.   Expert and resampled data for wet deposition rate (R in units of s^{-1}) corresponding to 0.3 μm particles and a rainfall intensity of 0.33 mm/10 min

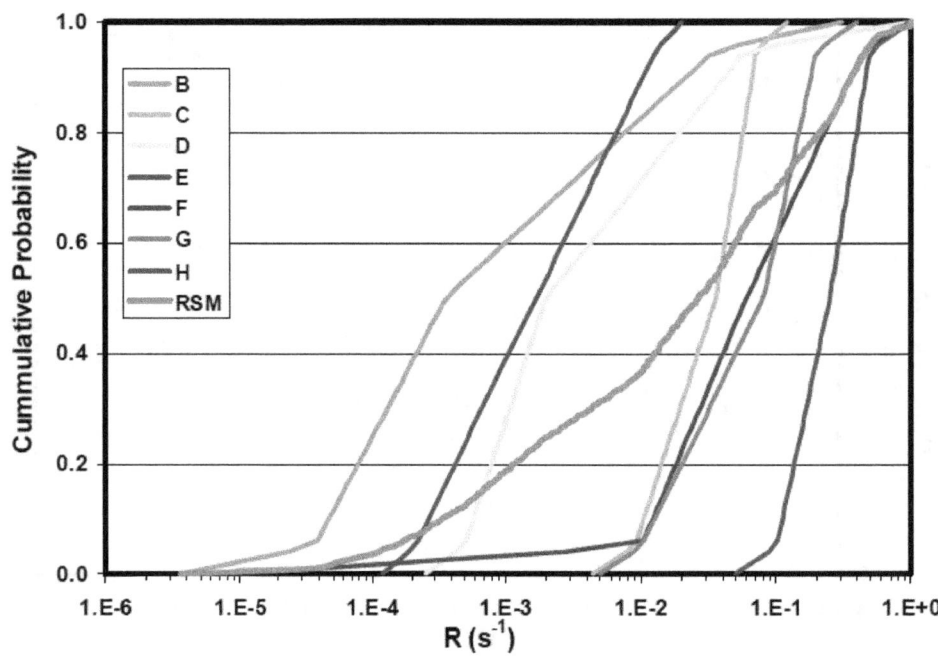

Figure 4-10.   Expert and resampled data for wet deposition rate (R in units of s^{-1}) corresponding to 0.3 μm particles and a rainfall intensity of 1.67 mm/10 min

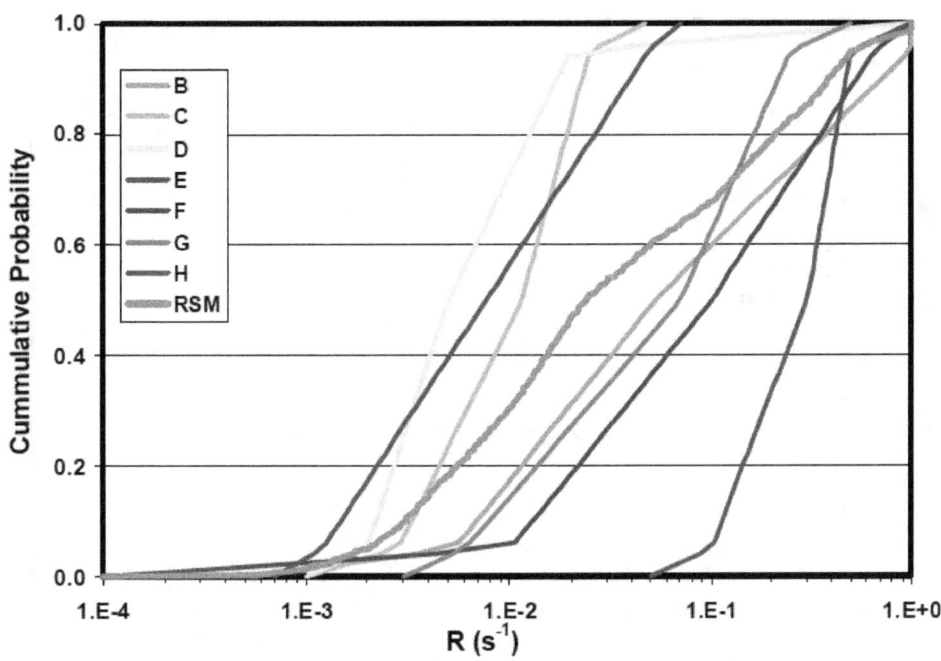

Figure 4-11. Expert and resampled data for wet deposition rate (R in units of s^{-1}) corresponding to 1 μm particles and a rainfall intensity of 0.3 mm/hr

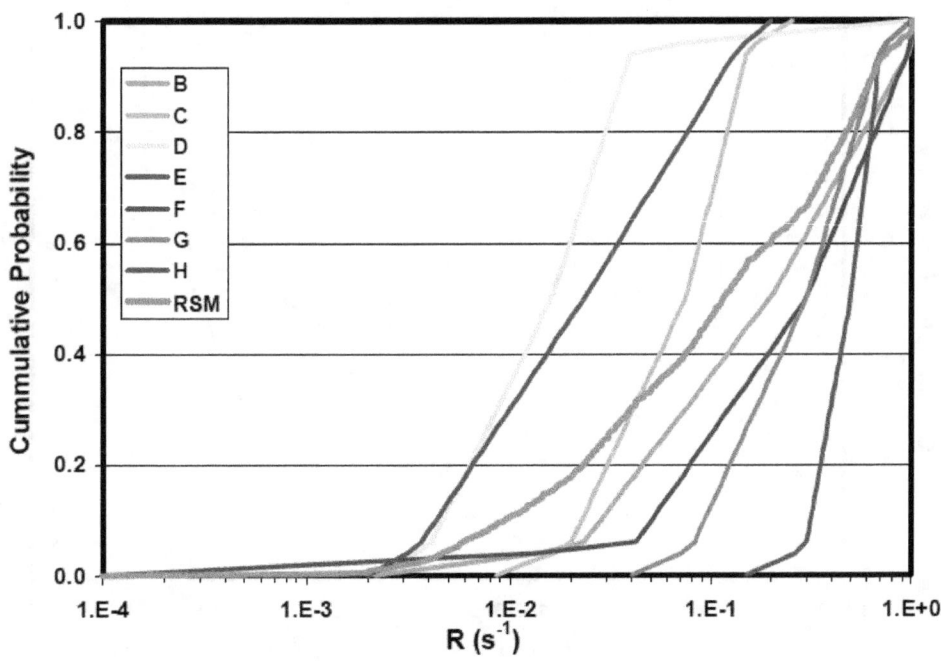

Figure 4-12. Expert and resampled data for wet deposition rate (R in units of s^{-1}) corresponding to 1 μm particles and a rainfall intensity of 2 mm/hr

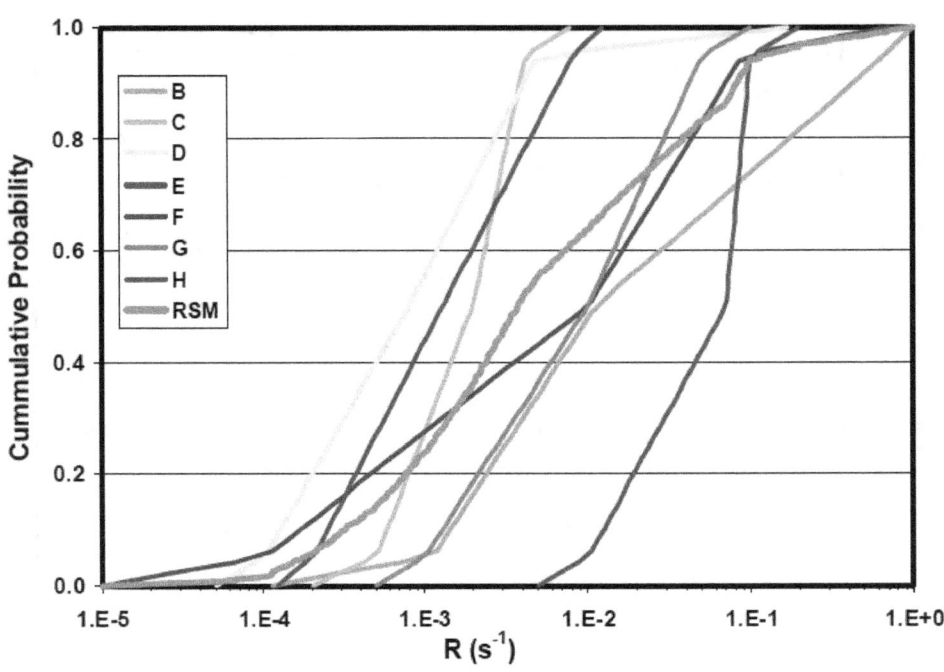

Figure 4-13. Expert and resampled data for wet deposition rate (R in units of s^{-1}) corresponding to 1 μm particles and a rainfall intensity of 0.05 mm/10 min

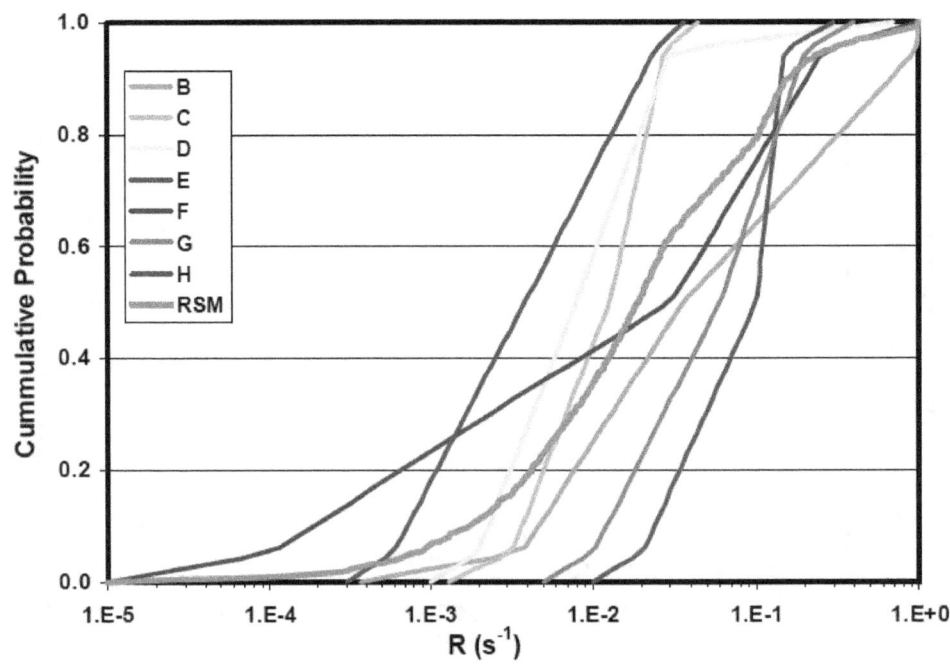

Figure 4-14. Expert and resampled data for wet deposition rate (R in units of s^{-1}) corresponding to 1 μm particles and a rainfall intensity of 0.33 mm/10 min

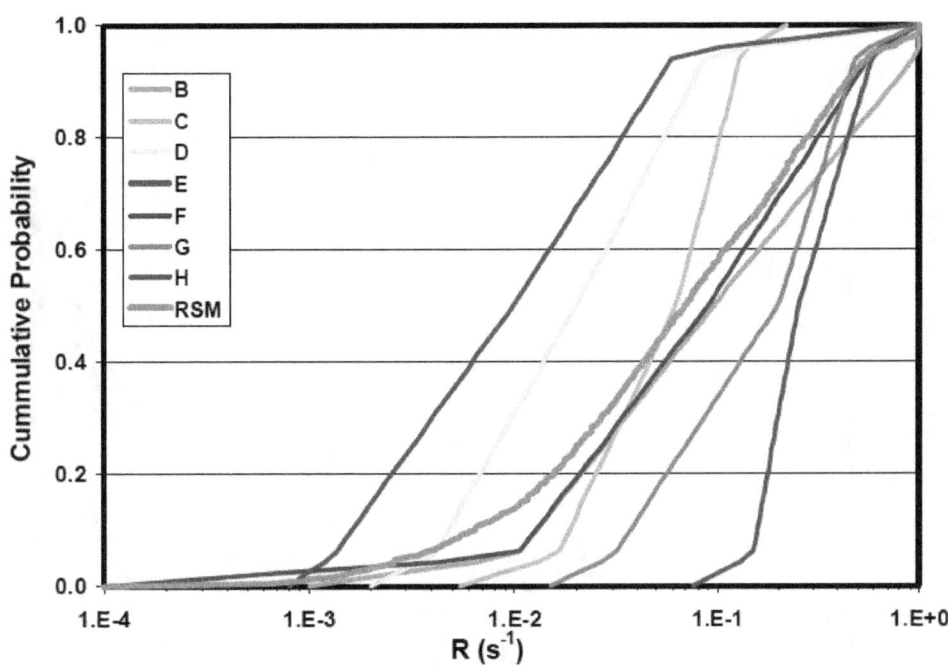

Figure 4-15. Expert and resampled data for wet deposition rate (R in units of s^{-1}) corresponding to 1 μm particles and a rainfall intensity of 1.67 mm/10 min

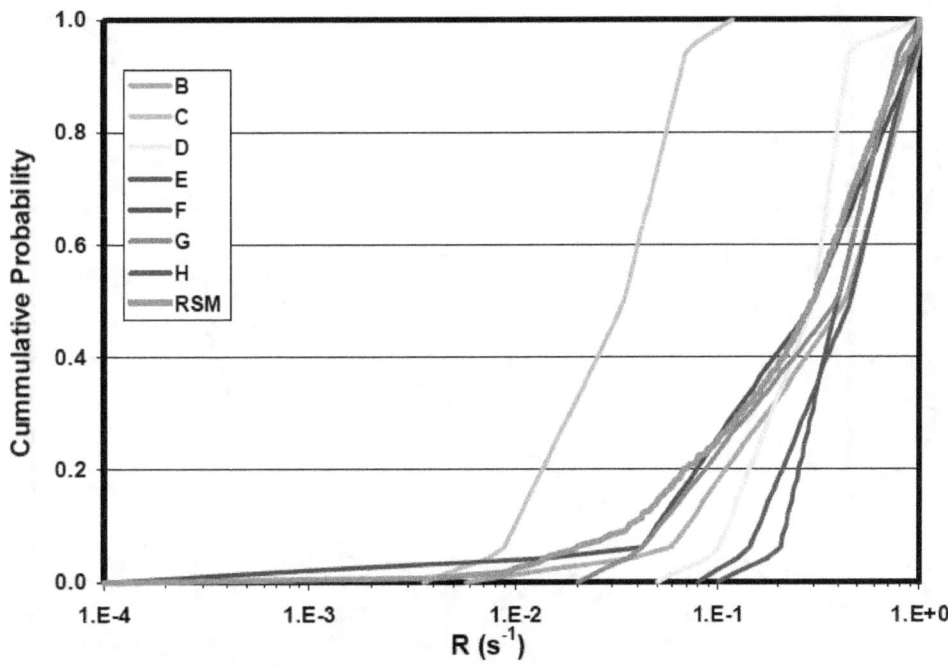

Figure 4-16. Expert and resampled data for wet deposition rate (R in units of s^{-1}) corresponding to 10 μm particles and a rainfall intensity of 0.3 mm/hr

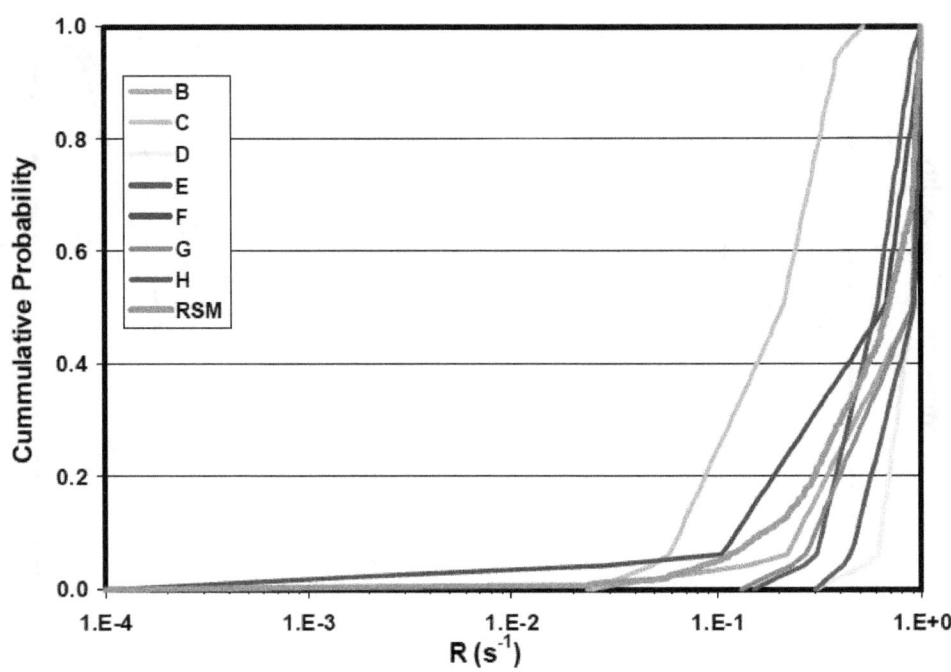

Figure 4-17. Expert and resampled data for wet deposition rate (R in units of s^{-1}) corresponding to 10 μm particles and a rainfall intensity of 2 mm/hr

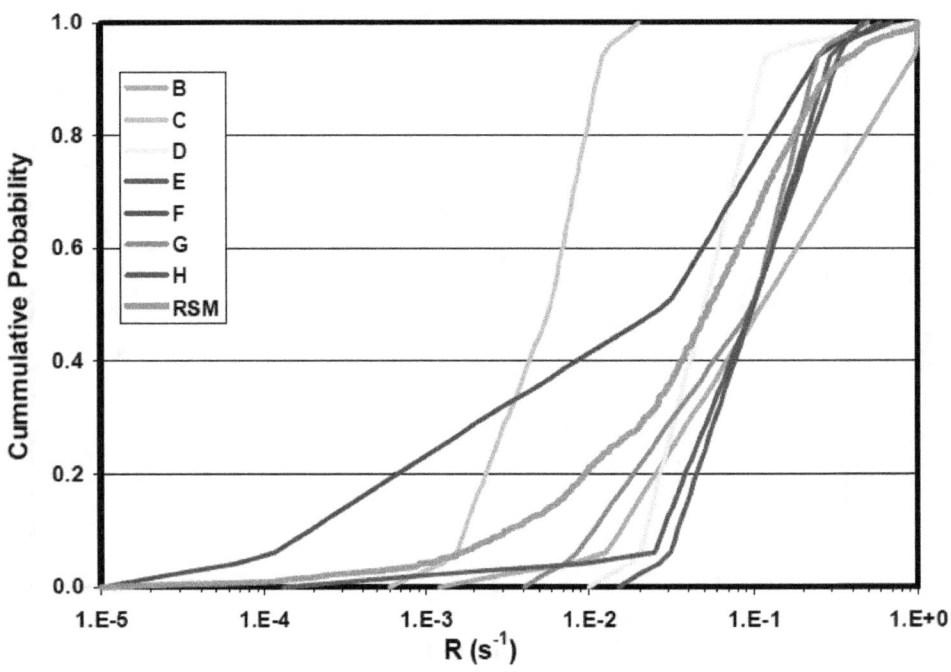

Figure 4-18. Expert and resampled data for wet deposition rate (R in units of s^{-1}) corresponding to 10 μm particles and a rainfall intensity of 0.05 mm/10 min

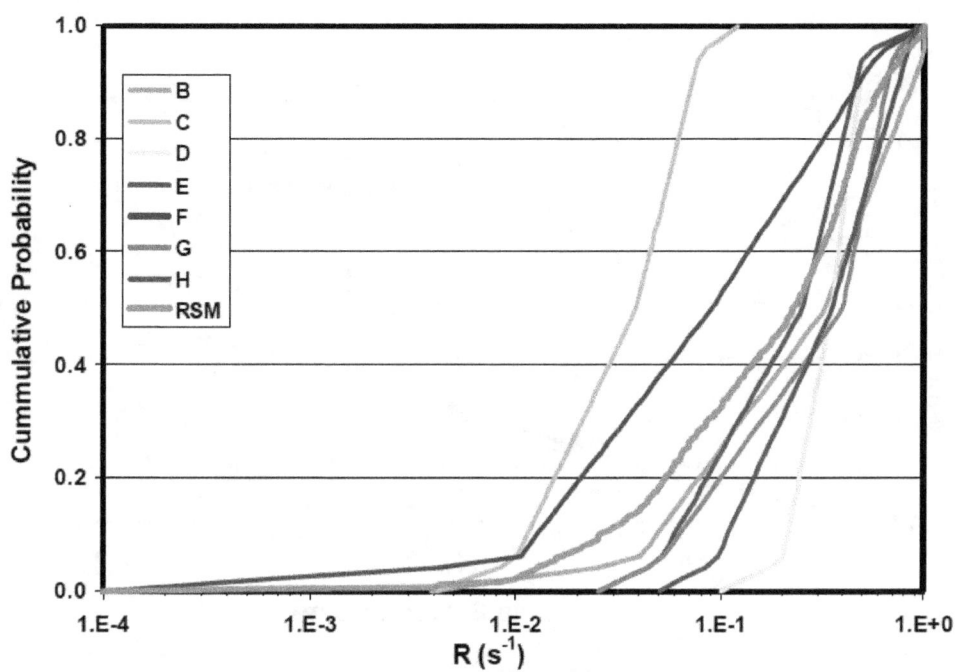

Figure 4-19. Expert and resampled data for wet deposition rate (R in units of s⁻¹) corresponding to 10 μm particles and a rainfall intensity of 0.33 mm/10 min

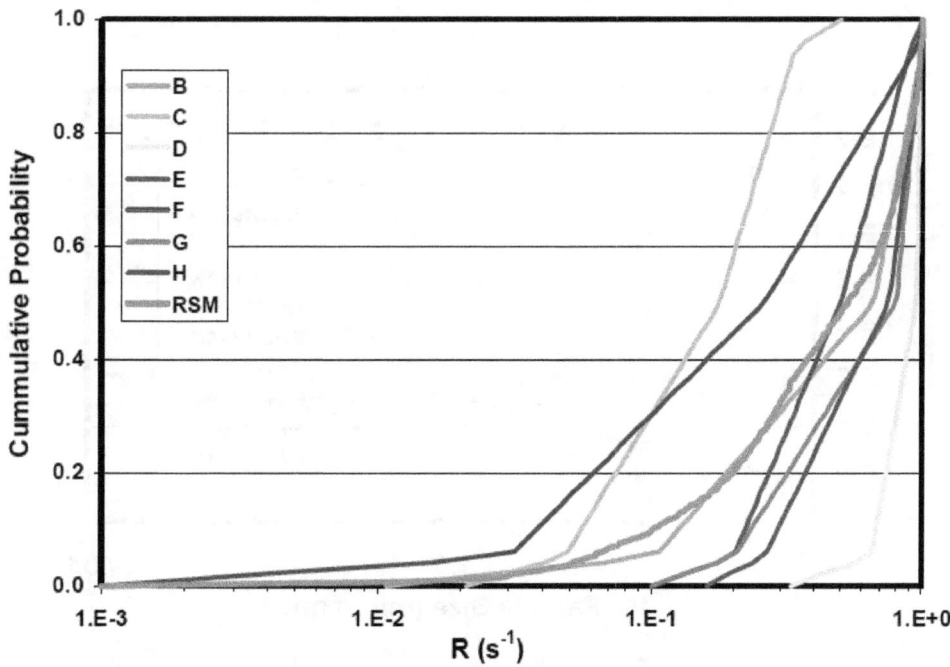

Figure 4-20. Expert and resampled data for wet deposition rate (R in units of s⁻¹) corresponding to 10 μm particles and a rainfall intensity of 1.67 mm/10 min

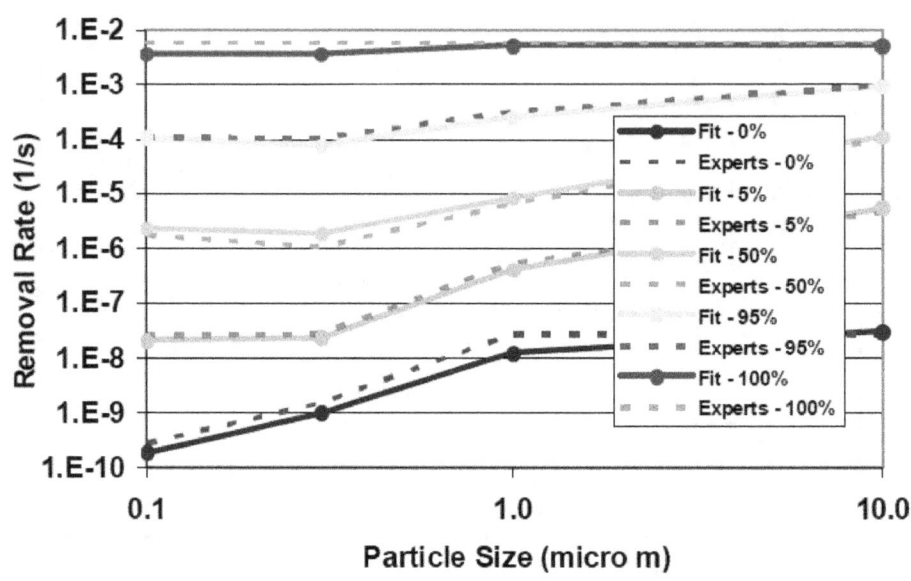

Figure 4-21. Correlated values (fit) and resampled expert data for removal rate at a rainfall intensity of 0.3 mm/hr as a function of particle size

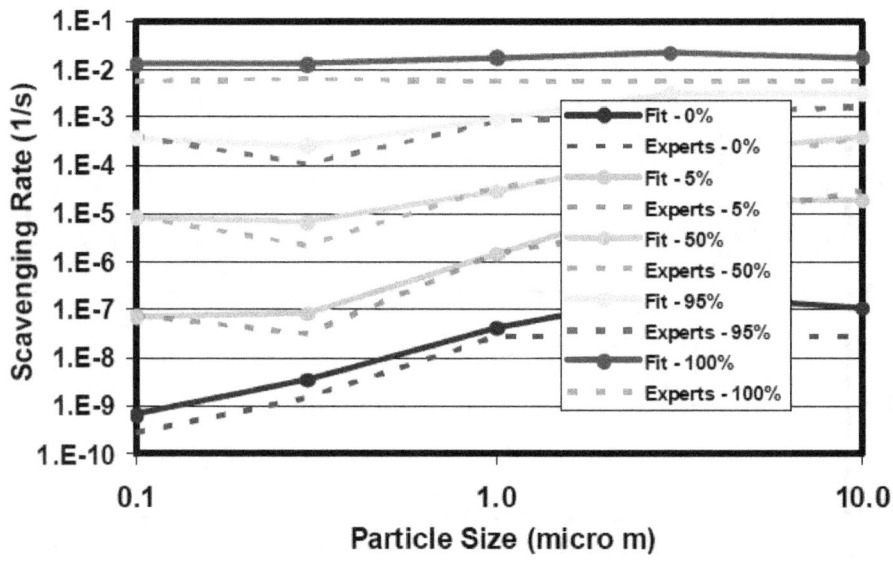

Figure 4-22. Correlated values (fit) and resampled expert data for removal rate at a rainfall intensity of 2 mm/hr as a function of particle size

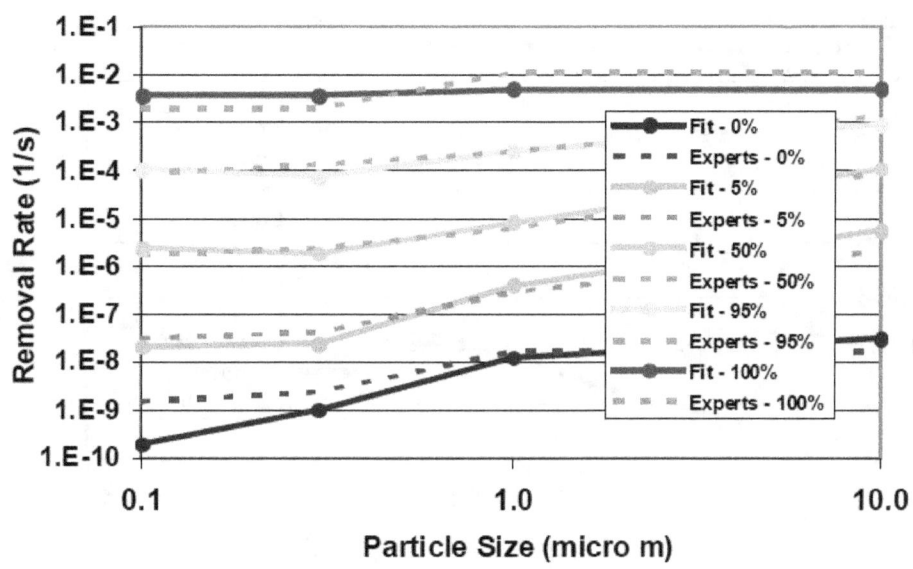

Figure 4-23. Correlated values (fit) and resampled expert data for removal rate at a rainfall intensity of 0.05 mm/10 min as a function of particle size

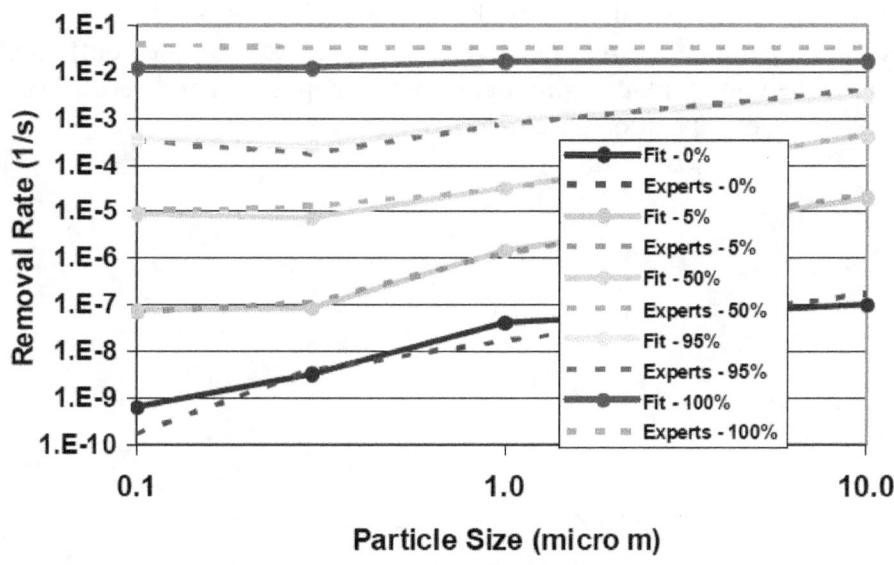

Figure 4-24. Correlated values (fit) and resampled expert data for removal rate at a rainfall intensity of 0.33 mm/10 min as a function of particle size

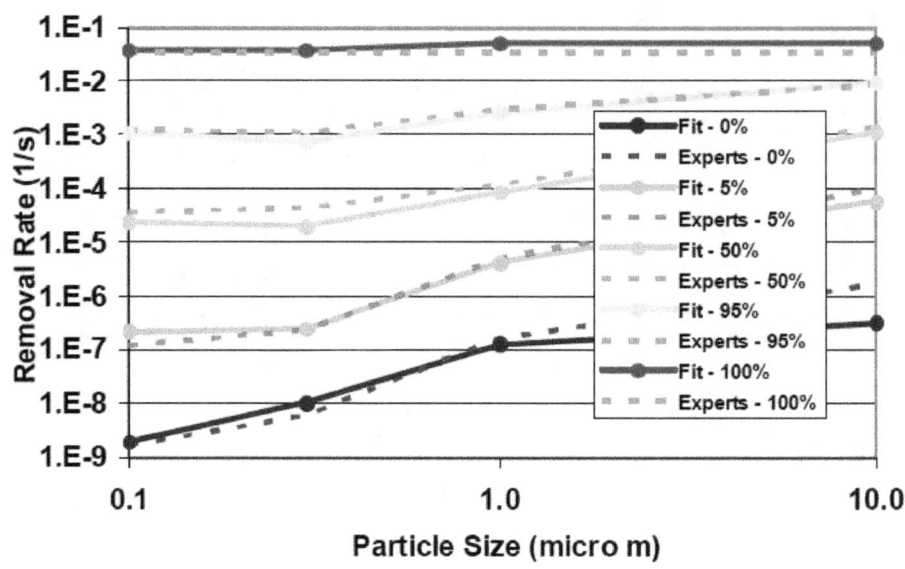

Figure 4-25.   Correlated values (fit) and resampled expert data for removal rate at a rain-fall intensity of 1.67 mm/10 min as a function of particle size

4.2   Correlations Between Wet Deposition Data

Because a single correlation is currently used to represent wet deposition for all aerosol bins in MACCS2, no correlation of the wet deposition values is needed. However, if a model were used in which wet deposition parameters were needed for each aerosol bin, the coefficients should be correlated to preserve the correct relationship between wet deposition and aerosol size. A correlation coefficient between wet deposition parameters of 0.9 to 1.0 is reasonable.

5.0 LATENT HEALTH EFFECTS

MACCS2 allows users to construct relationships between doses to an individual organ (or an effective dose to the whole body) and the risk of induction of a specific cancer related to that organ. The category, "other," captures the risk of other type of cancers not specifically included in the list. The category, "all cancers," is the total of all types of cancers induced by the radiation exposure. The constant of proportionality between dose and an induced health effect is called a risk factor. NUREG/CR-6555 [Reference 5] documents an expert elicitation of such risk factors. The data from this expert elicitation is evaluated in this section.

5.1   Distributions of Risk Factors for Late Health Effects

Table 5-1 provides a summary of the results. The values in the table correspond to the risk to an individual exposed to 1 Gy of low-LET radiation over a 1-min time period. This corresponds to a whole-body dose of 1 Sv (100 rem). The values in the table are population averaged, assuming an equal number of male and female members of the population.

Values are tabulated at 11 quantile levels and for the mean and mode. Figures 5-1 through 5-12 compare the aggregated distributions, using the resampling methodology, with the distributions provided by each of the experts for each of the 12 cancer categories.

Table 5-1:   Risk factors for latent cancer fatalities.

Quantile	Risk of a Latent Cancer Fatality (1/Gy)											
	Bone	Colon	Breast	Leuke-mia	Liver	Lung	Pancreas	Skin	Stomach	Thyroid	Other	All Cancers
0.00	0	$4.0 \cdot 10^{-5}$	$5.0 \cdot 10^{-5}$	$1.0 \cdot 10^{-4}$	0	$7.0 \cdot 10^{-5}$	0	0	0	0	0	$1.5 \cdot 10^{-2}$
0.01	0	$2.8 \cdot 10^{-4}$	$9.0 \cdot 10^{-4}$	$1.5 \cdot 10^{-3}$	$1.8 \cdot 10^{-5}$	$6.4 \cdot 10^{-4}$	0	0	$1.0 \cdot 10^{-4}$	$2.3 \cdot 10^{-5}$	$1.7 \cdot 10^{-3}$	$1.8 \cdot 10^{-2}$
0.05	$3.7 \cdot 10^{-5}$	$1.4 \cdot 10^{-3}$	$1.9 \cdot 10^{-3}$	$3.4 \cdot 10^{-3}$	$1.2 \cdot 10^{-4}$	$5.1 \cdot 10^{-3}$	$5.4 \cdot 10^{-5}$	$1.6 \cdot 10^{-5}$	$3.5 \cdot 10^{-4}$	$8.4 \cdot 10^{-5}$	$5.7 \cdot 10^{-3}$	$3.4 \cdot 10^{-2}$
0.10	$6.2 \cdot 10^{-5}$	$2.5 \cdot 10^{-3}$	$2.5 \cdot 10^{-3}$	$4.5 \cdot 10^{-3}$	$2.1 \cdot 10^{-4}$	$7.2 \cdot 10^{-3}$	$1.5 \cdot 10^{-4}$	$5.1 \cdot 10^{-5}$	$6.3 \cdot 10^{-4}$	$1.4 \cdot 10^{-4}$	$8.5 \cdot 10^{-3}$	$4.2 \cdot 10^{-2}$
0.25	$1.2 \cdot 10^{-4}$	$5.3 \cdot 10^{-3}$	$4.5 \cdot 10^{-3}$	$6.6 \cdot 10^{-3}$	$4.1 \cdot 10^{-4}$	$1.4 \cdot 10^{-2}$	$5.8 \cdot 10^{-4}$	$1.9 \cdot 10^{-4}$	$1.4 \cdot 10^{-3}$	$3.0 \cdot 10^{-4}$	$1.5 \cdot 10^{-2}$	$6.2 \cdot 10^{-2}$
0.50	$4.2 \cdot 10^{-4}$	$1.0 \cdot 10^{-2}$	$9.2 \cdot 10^{-3}$	$9.5 \cdot 10^{-3}$	$8.6 \cdot 10^{-4}$	$2.6 \cdot 10^{-2}$	$2.1 \cdot 10^{-3}$	$4.5 \cdot 10^{-4}$	$3.2 \cdot 10^{-3}$	$6.3 \cdot 10^{-4}$	$2.7 \cdot 10^{-2}$	$9.6 \cdot 10^{-2}$
0.75	$1.2 \cdot 10^{-3}$	$1.8 \cdot 10^{-2}$	$2.0 \cdot 10^{-2}$	$1.4 \cdot 10^{-2}$	$2.3 \cdot 10^{-3}$	$4.4 \cdot 10^{-2}$	$5.8 \cdot 10^{-3}$	$9.7 \cdot 10^{-4}$	$7.3 \cdot 10^{-3}$	$1.8 \cdot 10^{-3}$	$5.3 \cdot 10^{-2}$	$1.6 \cdot 10^{-1}$
0.90	$6.1 \cdot 10^{-3}$	$2.8 \cdot 10^{-2}$	$3.6 \cdot 10^{-2}$	$1.8 \cdot 10^{-2}$	$1.1 \cdot 10^{-2}$	$6.6 \cdot 10^{-2}$	$9.1 \cdot 10^{-3}$	$2.3 \cdot 10^{-3}$	$1.9 \cdot 10^{-2}$	$5.9 \cdot 10^{-3}$	$9.0 \cdot 10^{-2}$	$2.3 \cdot 10^{-1}$
0.95	$1.0 \cdot 10^{-2}$	$3.5 \cdot 10^{-2}$	$4.7 \cdot 10^{-2}$	$2.5 \cdot 10^{-2}$	$1.9 \cdot 10^{-2}$	$8.4 \cdot 10^{-2}$	$1.2 \cdot 10^{-2}$	$3.2 \cdot 10^{-3}$	$3.5 \cdot 10^{-2}$	$1.3 \cdot 10^{-2}$	$1.2 \cdot 10^{-1}$	$2.9 \cdot 10^{-1}$
0.99	$1.9 \cdot 10^{-2}$	$5.8 \cdot 10^{-2}$	$6.1 \cdot 10^{-2}$	$3.9 \cdot 10^{-2}$	$3.6 \cdot 10^{-2}$	$1.3 \cdot 10^{-1}$	$1.9 \cdot 10^{-2}$	$5.5 \cdot 10^{-3}$	$7.1 \cdot 10^{-2}$	$5.6 \cdot 10^{-2}$	$2.0 \cdot 10^{-1}$	$4.9 \cdot 10^{-1}$
1.00	$3.0 \cdot 10^{-2}$	$8.0 \cdot 10^{-2}$	$1.0 \cdot 10^{-1}$	$7.0 \cdot 10^{-2}$	$7.6 \cdot 10^{-2}$	$2.0 \cdot 10^{-1}$	$2.8 \cdot 10^{-2}$	$1.0 \cdot 10^{-2}$	$1.5 \cdot 10^{-1}$	$1.2 \cdot 10^{-1}$	$2.8 \cdot 10^{-1}$	$7.0 \cdot 10^{-1}$
Mean	$1.9 \cdot 10^{-3}$	$1.3 \cdot 10^{-2}$	$1.5 \cdot 10^{-2}$	$1.1 \cdot 10^{-2}$	$3.8 \cdot 10^{-3}$	$3.3 \cdot 10^{-2}$	$3.7 \cdot 10^{-3}$	$8.7 \cdot 10^{-4}$	$8.2 \cdot 10^{-3}$	$3.5 \cdot 10^{-3}$	$4.1 \cdot 10^{-2}$	$1.2 \cdot 10^{-1}$
Mode	$6.1 \cdot 10^{-5}$	$4.7 \cdot 10^{-3}$	$3.1 \cdot 10^{-3}$	$8.0 \cdot 10^{-3}$	$3.9 \cdot 10^{-4}$	$1.2 \cdot 10^{-2}$	$1.5 \cdot 10^{-4}$	$1.4 \cdot 10^{-5}$	$8.6 \cdot 10^{-4}$	$2.1 \cdot 10^{-4}$	$1.6 \cdot 10^{-2}$	$7.2 \cdot 10^{-2}$

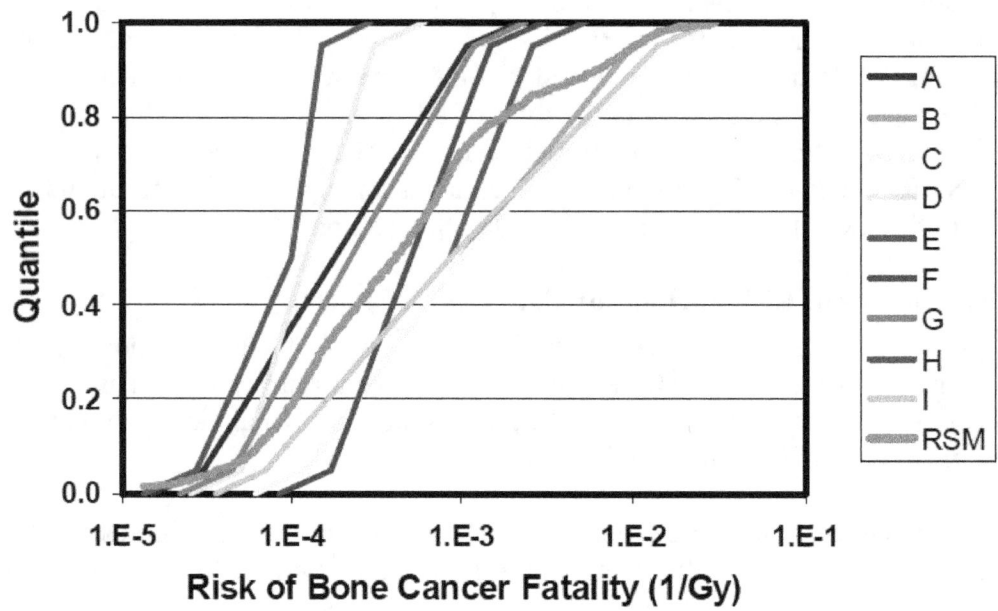

Figure 5-1.   Expert and resampled data for the risk factor for a bone cancer fatality to a person receiving 1 Gy low-LET radiation over 1 min

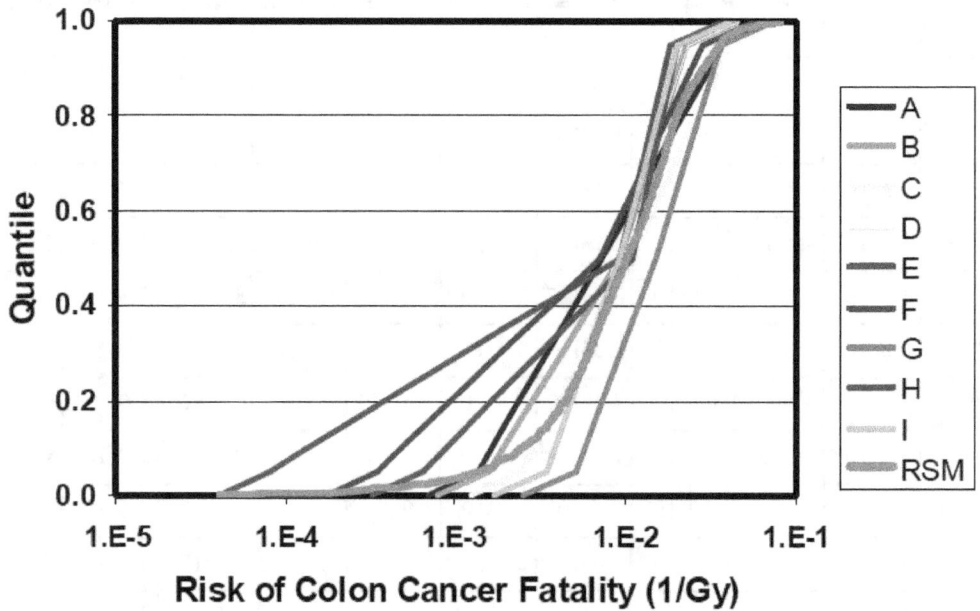

Figure 5-2.   Expert and resampled data for the risk factor for a colon cancer fatality to a person receiving 1 Gy low-LET radiation over 1 min

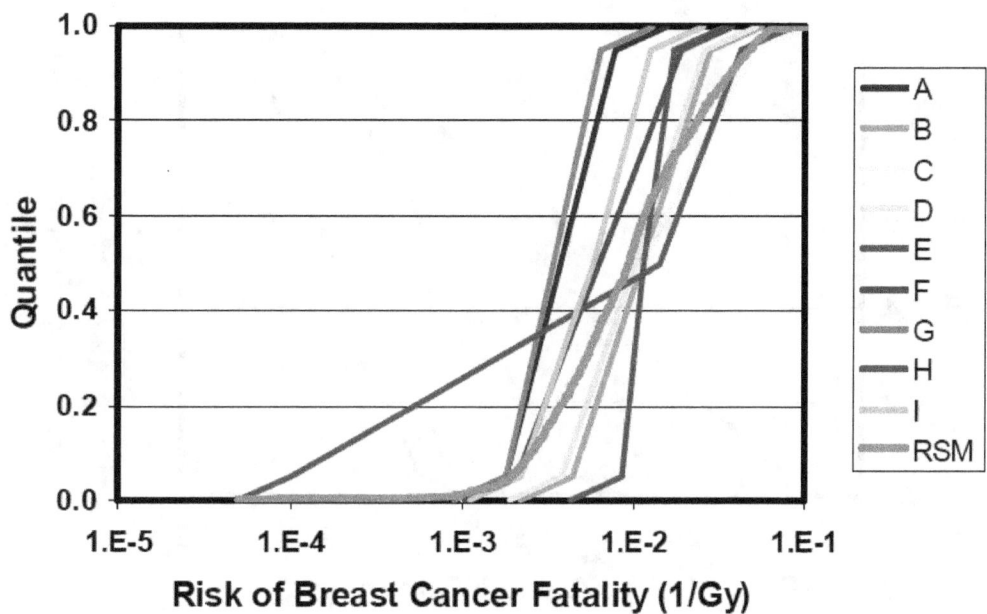

Figure 5-3. Expert and resampled data for the risk factor for a breast cancer fatality to a person receiving 1 Gy low-LET radiation over 1 min

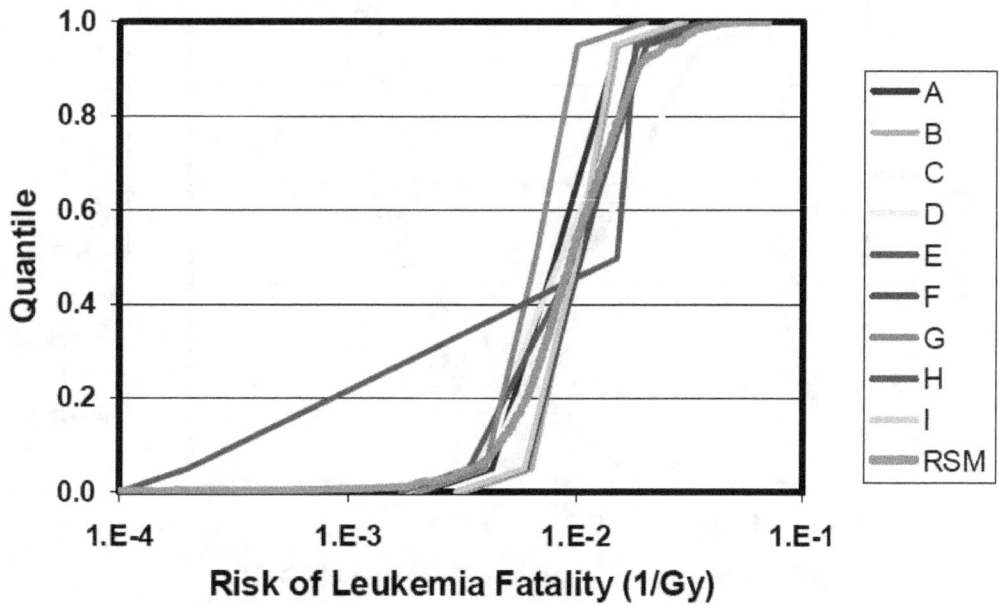

Figure 5-4. Expert and resampled data for the risk factor for a leukemia fatality to a person receiving 1 Gy low-LET radiation over 1 min

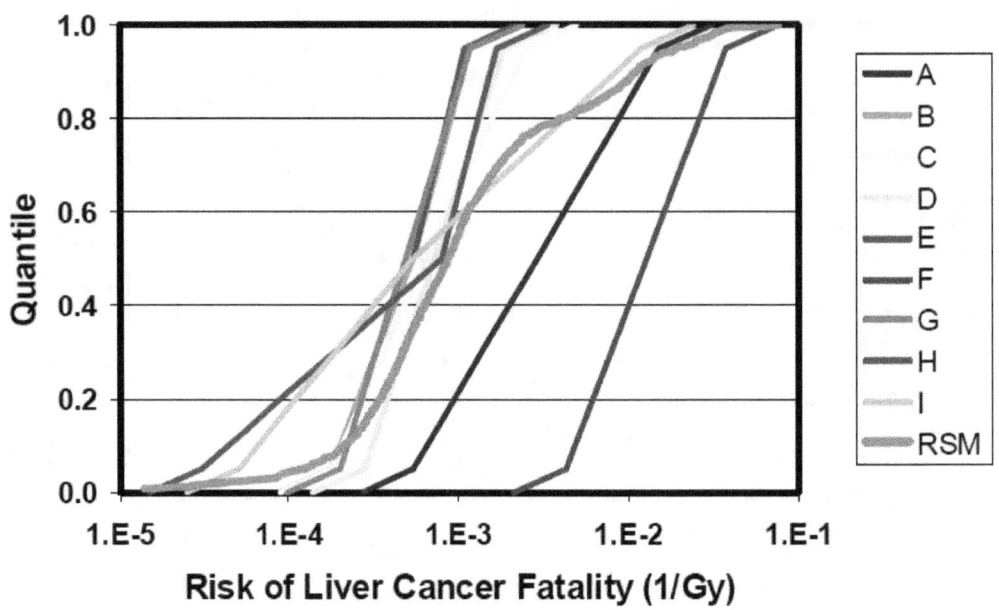

Figure 5-5. Expert and resampled data for the risk factor for a liver cancer fatality to a person receiving 1 Gy low-LET radiation over 1 min

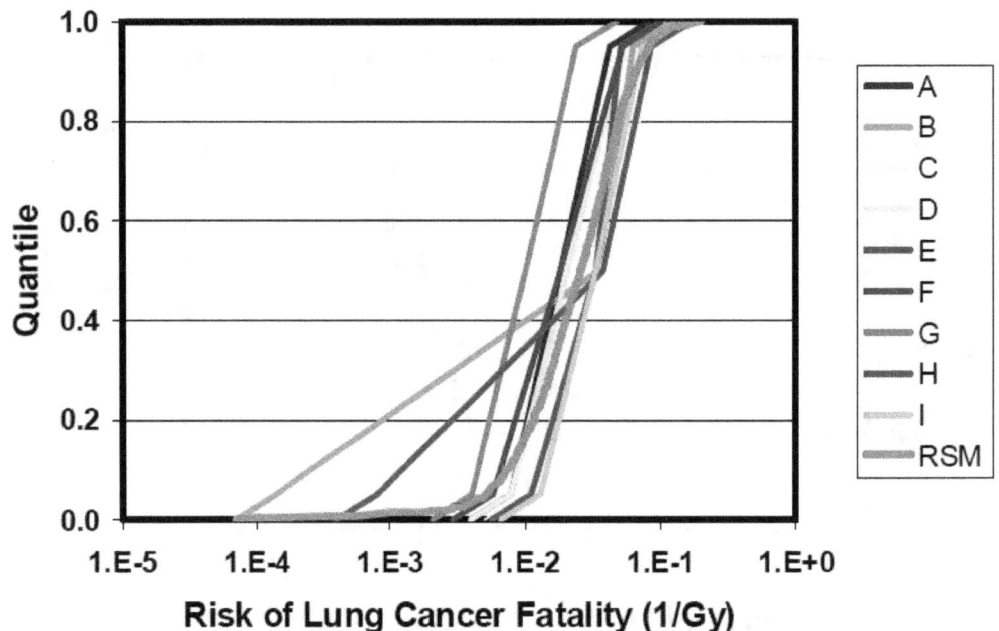

Figure 5-6. Expert and resampled data for the risk factor for a lung cancer fatality to a person receiving 1 Gy low-LET radiation over 1 min

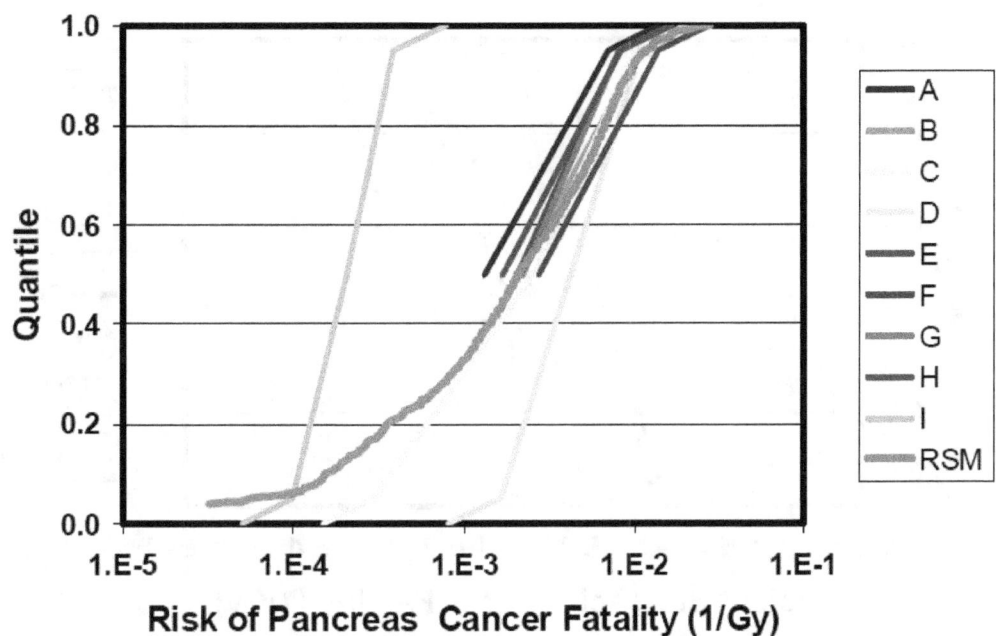

Figure 5-7. Expert (excluding values of zero) and resampled data for the risk factor for a pancreas cancer fatality to a person receiving 1 Gy low-LET radiation over 1 min

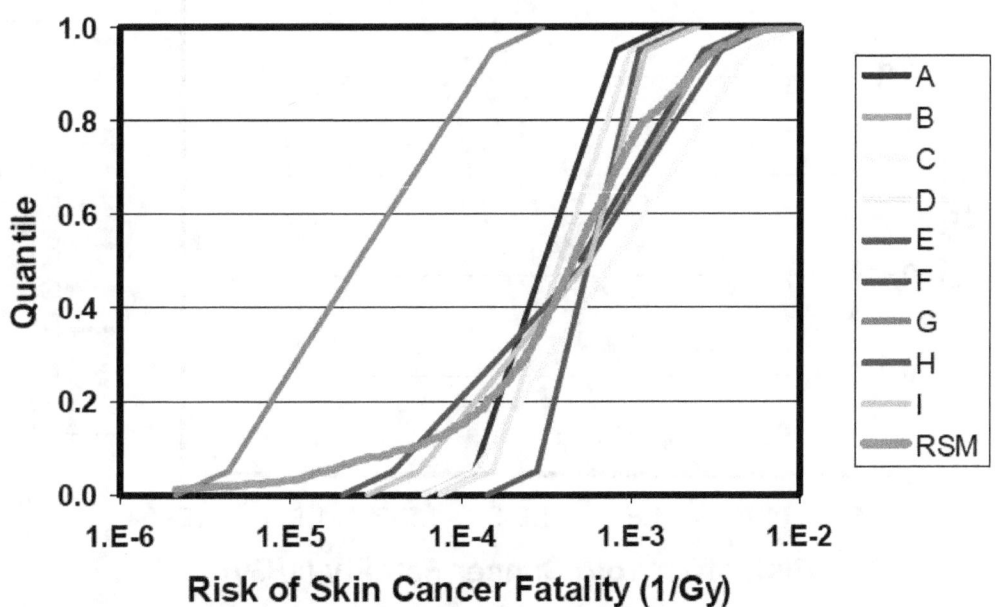

Figure 5-8. Expert and resampled data for the risk factor for a skin cancer fatality to a person receiving 1 Gy low-LET radiation over 1 min

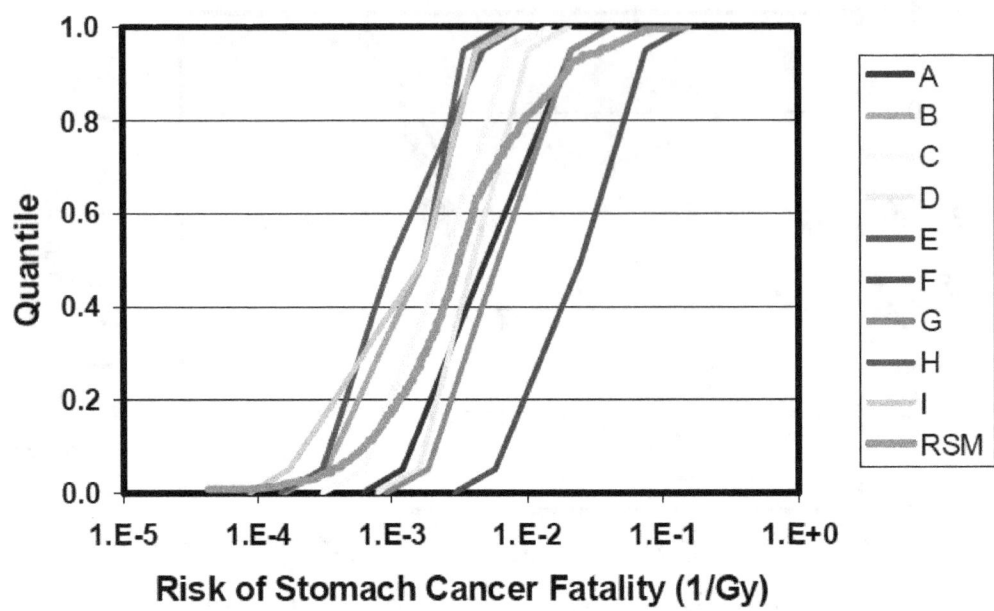

Figure 5-9. Expert and resampled data for the risk factor for a stomach cancer fatality to a person receiving 1 Gy low-LET radiation over 1 min

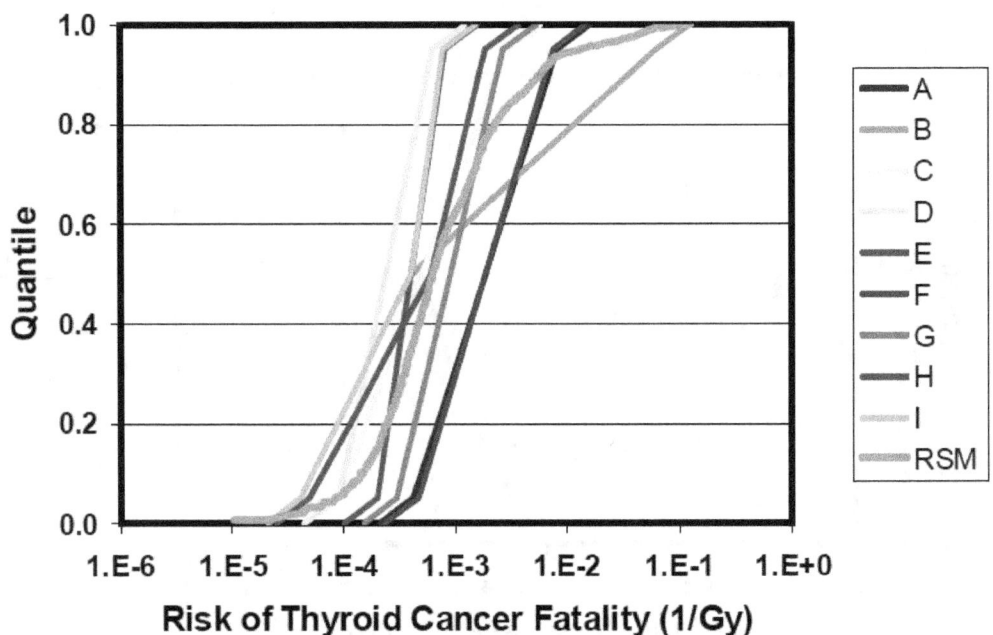

Figure 5-10. Expert and resampled data for the risk factor for a thyroid cancer fatality to a person receiving 1 Gy low-LET radiation over 1 min

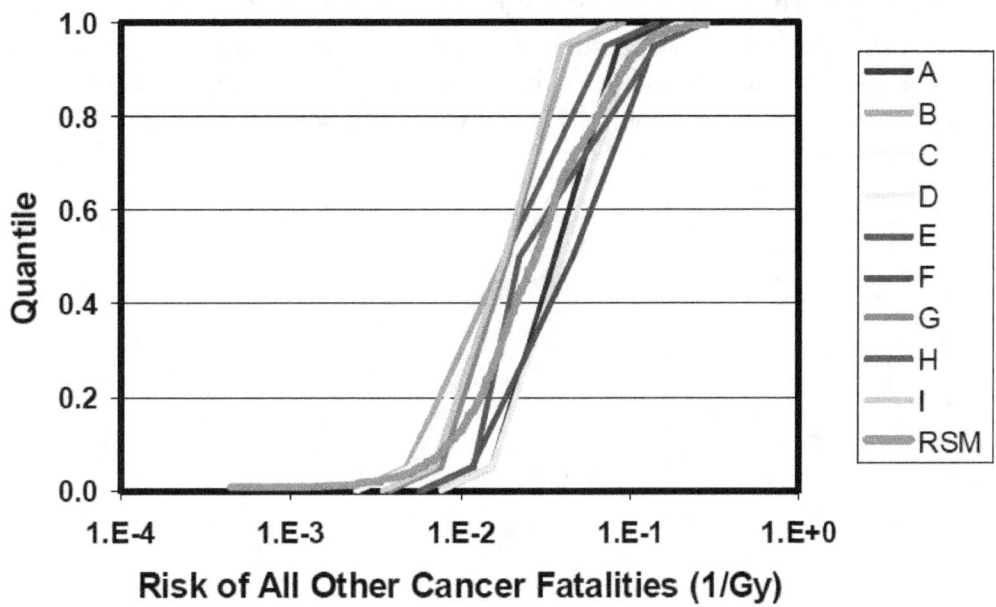

Figure 5-11. Expert and resampled data for the risk factor for other cancer fatality to a person receiving 1 Gy low-LET radiation over 1 min

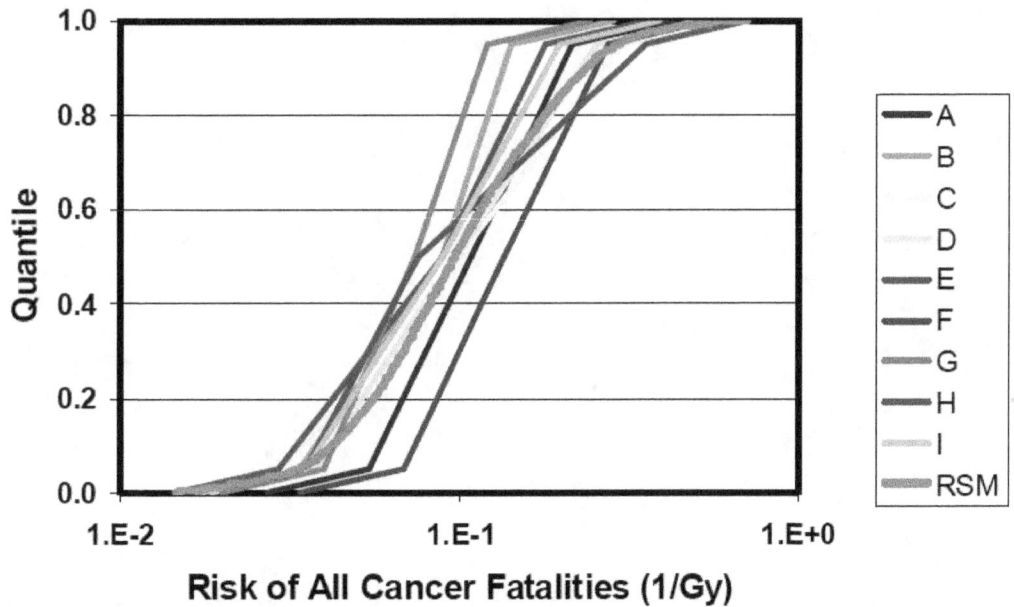

Figure 5-12. Expert and resampled data for the risk factor for all cancer fatalities to a person receiving 1 Gy low-LET radiation over 1 min

5.2 Correlations Between Risk Factors for Late Health Effects

It is not obvious that the health risks associated with doses to the human organs are interrelated. Cancer incidents and fatalities are estimated independently for each type of cancer. Therefore, the risk factors provided in Table 5-1 should be treated as being uncorrelated.

6.0 EARLY HEALTH EFFECTS

6.1 Distributions of Early Health Parameters

MACCS2 allows users to construct relationships between acute doses to an individual organ and the risk of induction of an early injury or fatality related to that organ. The relationship between dose and an induced health effect is represented by a Weibull function, which has three parameters: a threshold, a D_{50} or LD_{50} value, and a shape factor, β. The form of this equation is as follows:

$$R = 1 - \exp\left(-\ln(2) \cdot \left(\frac{D}{D_{50}}\right)^{\beta}\right) \tag{6.1}$$

NUREG/CR-6545 [Reference 4] documents an expert elicitation of the relationship between dose and induced early health effects. The data from this expert elicitation are evaluated in this section.

Four specific health effects are considered in this report. Table 6-1 provides a summary of the results. The values in Table 6-1 for a given health effect should be treated as having a correlation coefficient of 1. That is, the same quantile values should be used for all three parameters associated with a health effect. Furthermore, the parameters associated with the pulmonary syndrome and pneumonitis should be treated as having a correlation coefficient of 1 because it would not make sense for pneumonitis to require higher doses than the pulmonary syndrome.

The first health effect considered is hematopoietic syndrome, a type of early fatality. The expert data are for a whole-body dose rate of 100 Gy/hr. Following the radiation exposure, the person is assumed to receive supportive medical treatment.

The second health effect considered is gastrointestinal syndrome, also a type of early fatality. The expert data are for a whole-body dose rate of 100 Gy/hr. Following the radiation exposure, the person is assumed to receive supportive medical treatment.

The third health effect considered is pulmonary syndrome, also a type of early fatality. The expert data are for a lung dose rate of 1 Gy/hr of beta radiation to a person under the age of 40. Supportive medical treatment is assumed following the exposure.

The fourth health effect considered is pneumonitis, an early health effect that is generally nonfatal. The expert data are for a lung dose rate of 1 Gy/hr of beta radiation to a person under the age of 40. Again, supportive medical treatment is assumed following the exposure.

Two of the three parameters in the Weibull function were elicited directly: dose threshold and D_{50} or LD_{50}. The other elicited values were D_{10} and D_{90}. The shape factor is calculated using the following equation:

$$\beta = \ln\left(\frac{\ln(0.5)}{\ln(0.9)}\right) / \ln\left(\frac{D50}{D10}\right) \tag{6.2}$$

Values are tabulated in Table 6-1 at 11 quantile levels and for the mean and mode of the distribution. Figures 6-1 through 6-8 compare the aggregated distributions, using the resampling methodology, with the distributions provided by each of the experts for each of the 4 early health effects.

Table 6-1: Coefficients for quantifying early health effects.

Quantile	Weibull Factors for Early Health Effects											
	Hematopoietic Syndrome			Gastrointestinal Syndrome			Pulmonary Syndrome			Pneumonitis		
	LD_{th} (Sv)	LD_{50} (Sv)	β	LD_{th} (Sv)	LD_{50} (Sv)	β	LD_{th} (Sv)	LD_{50} (Sv)	β	D_{th} (Sv)	D_{50} (Sv)	β
0.00	0.67	2.0	2.4	2.0	4.8	3.2	5.3	10.0	3.7	2.7	5.0	3.5
0.01	0.80	2.4	2.5	2.9	6.2	3.2	6.7	12.0	3.8	3.5	7.3	3.6
0.05	1.11	3.3	2.8	3.8	7.9	3.4	8.6	16.6	4.4	4.4	8.9	4.0
0.10	1.32	3.7	3.2	4.5	8.5	3.6	9.6	17.8	4.7	5.1	10.3	4.6
0.25	1.72	4.4	4.2	5.4	10.0	6.0	11.5	19.9	5.6	6.5	12.9	5.1
0.50	2.32	5.6	6.1	6.5	12.1	9.3	13.6	23.5	9.6	9.2	16.6	7.3
0.75	3.56	7.2	10.2	7.7	14.9	11.0	18.4	33.6	13.8	11.3	20.3	14.8
0.90	4.63	8.9	13.2	8.8	17.7	16.0	22.1	42.0	16.9	14.3	25.7	19.3
0.95	5.26	10.3	14.3	9.5	19.1	18.0	24.0	45.0	18.7	16.6	31.1	22.1
0.99	6.19	11.8	15.8	13.0	23.3	19.5	32.4	55.7	21.4	20.7	36.5	65.4
1.00	8.55	16.5	16.0	15.0	30.0	19.9	37.5	76.5	21.7	28.5	55.5	83.8
Mean	2.70	6.0	7.4	6.6	12.7	9.3	15.1	27.3	10.1	9.5	17.5	11.1
Mode	2.18	5.6	6.0	6.5	10.3	4.4	12.0	20.1	4.8	6.6	16.4	4.7

It is not entirely obvious how to best interpolate the data in Table 6-1 between quantile levels. Since all of the quantities in the table are positive, a logarithmic interpolation can be used and is likely to most faithfully reproduce the intention of the experts.

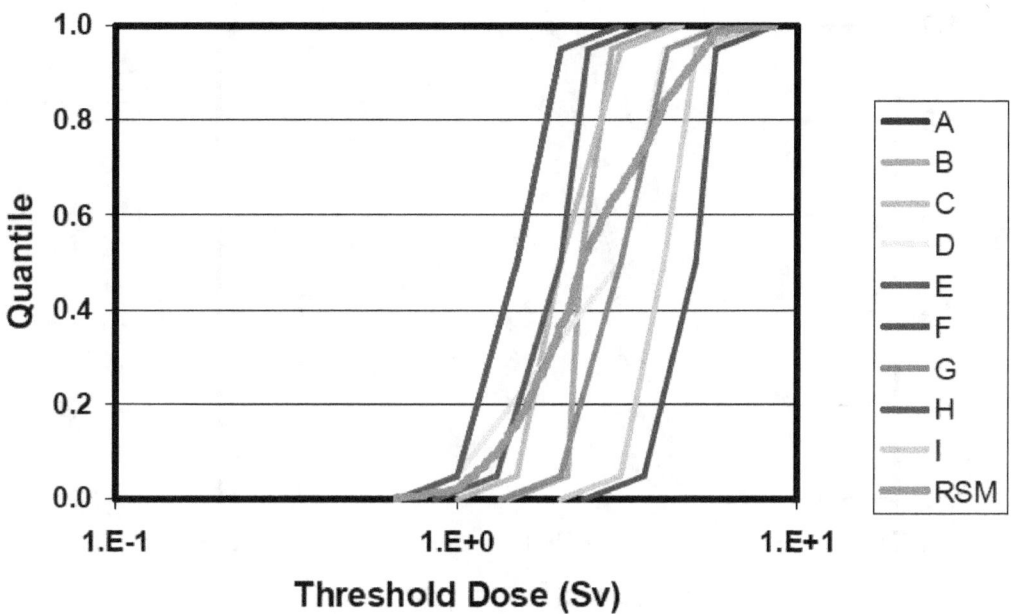

Figure 6-1. Expert and resampled data for the threshold dose corresponding to hematopoietic syndrome for a person receiving 100 Gy/hr low-LET radiation to the whole body

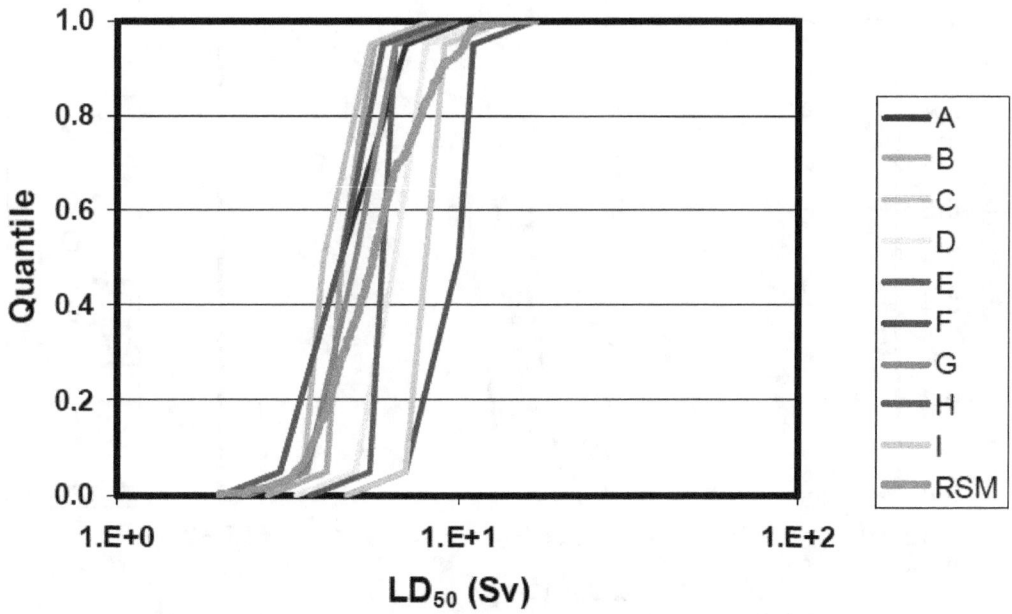

Figure 6-2. Expert and resampled data for LD_{50} corresponding to hematopoietic syndrome for a person receiving 100 Gy/hr low-LET radiation to the whole body

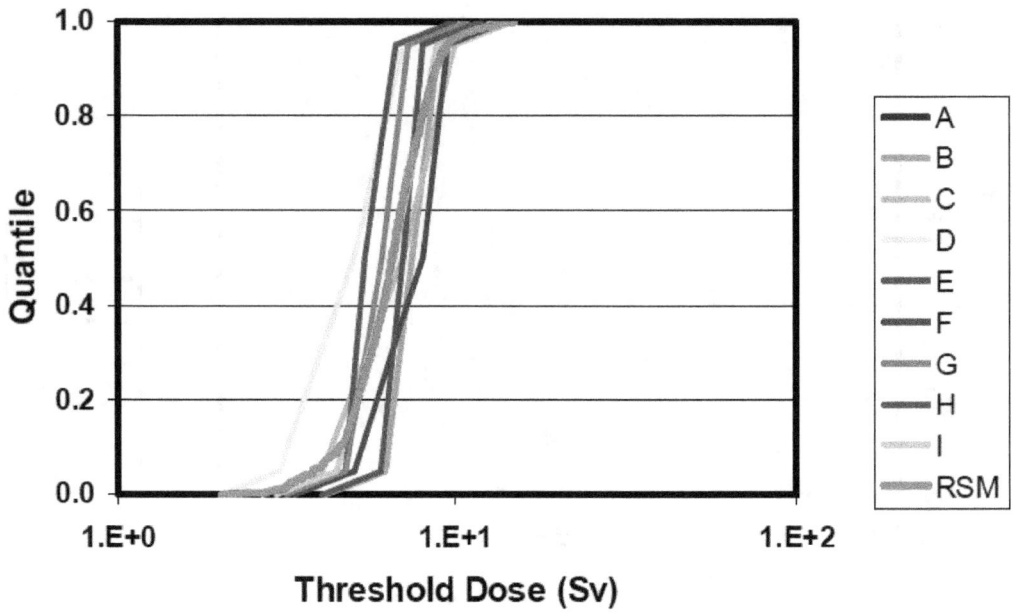

Figure 6-3.   Expert and resampled data for the threshold dose corresponding to gastrointestinal syndrome for a person receiving 100 Gy/hr low-LET radiation to the whole body

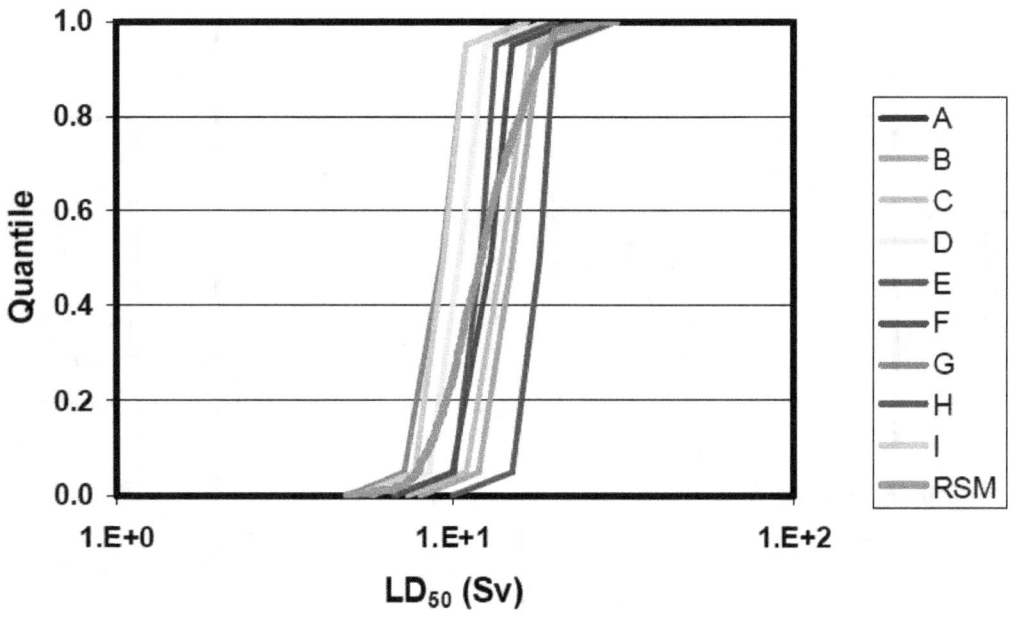

Figure 6-4.   Expert and resampled data for LD_{50} corresponding to gastrointestinal syndrome for a person receiving 100 Gy/hr low-LET radiation to the whole body

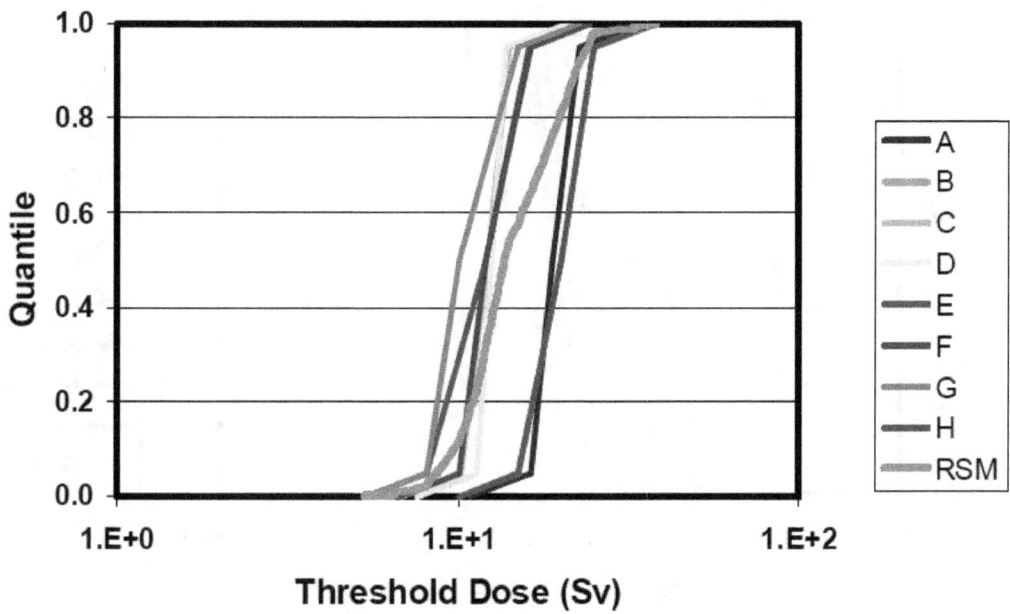

Figure 6-5. Expert and resampled data for the threshold dose corresponding to pulmonary syndrome for a person receiving 1 Gy/hr beta radiation to the lungs

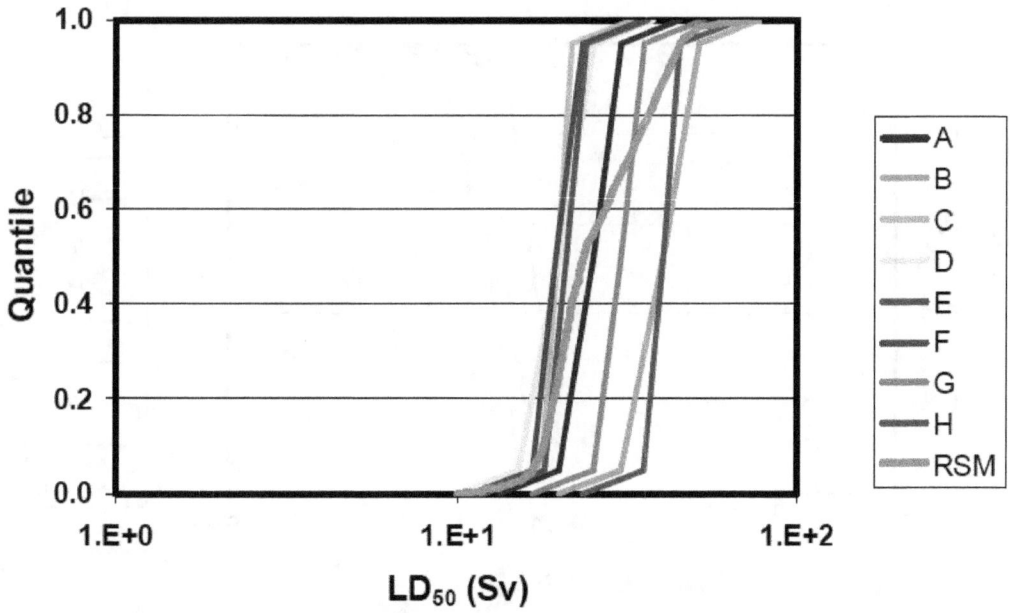

Figure 6-6. Expert and resampled data for LD_{50} corresponding to pulmonary syndrome for a person receiving 1 Gy/hr beta radiation to the lungs

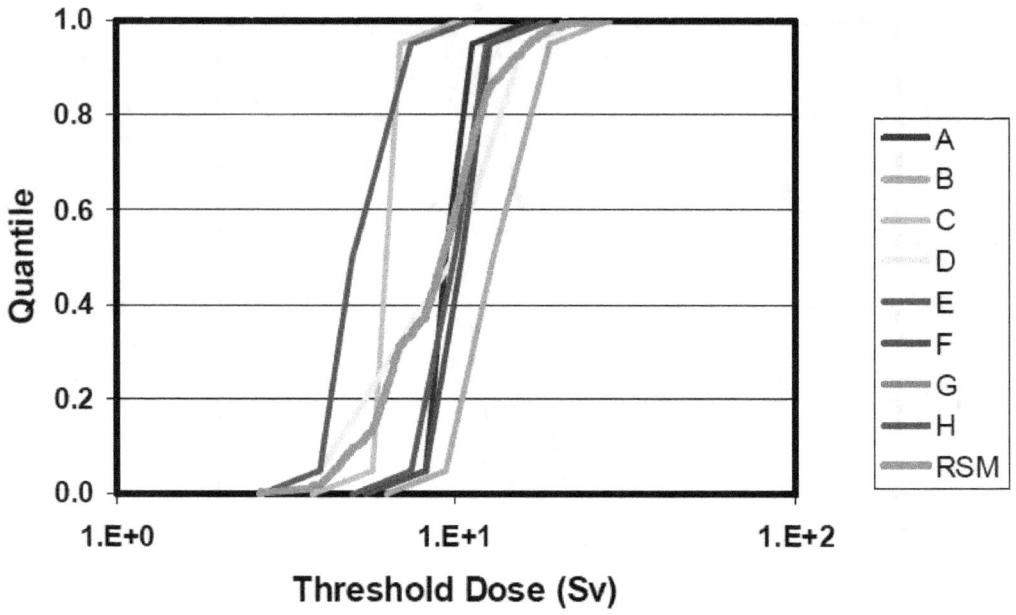

Figure 6-7. Expert and resampled data for the threshold dose corresponding to pneumonitis for a person receiving 1 Gy/hr beta radiation to the lungs

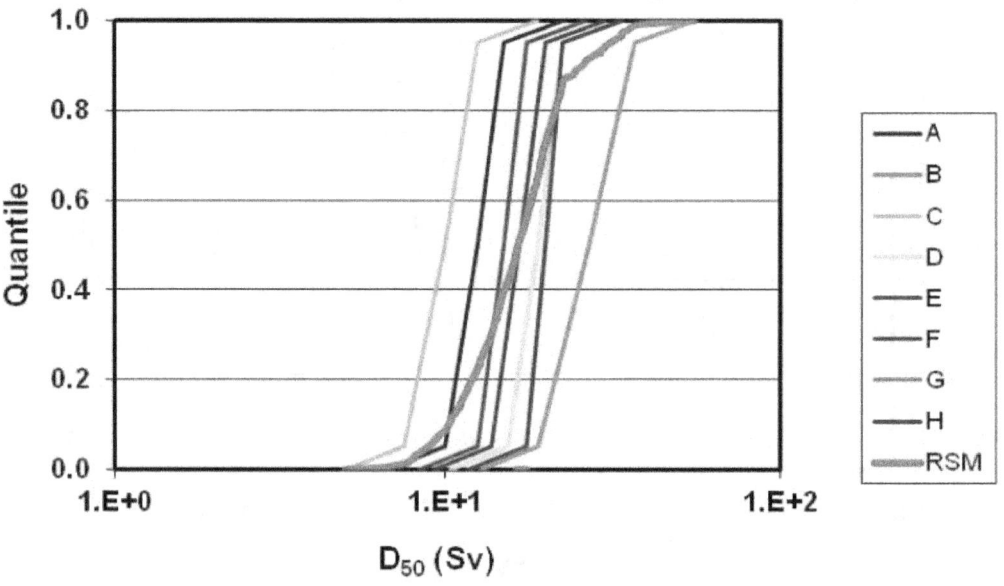

Figure 6-8. Expert and resampled data for D_{50} corresponding to pneumonitis for a person receiving 1 Gy/hr beta radiation to the lungs

6.2   Correlations Between Early Health Parameters

It is not obvious that the acute health risks associated with doses to the human organs are interrelated. Therefore, the risk factors provided in Table 6-1 should be treated as being uncorrelated for different organs. However, the values in Table 6-1 for a given health effect should be treated as having a correlation coefficient of 1. That is, the same quantile values should be used for all three parameters associated with a health effect. Furthermore, the parameters associated with the pulmonary syndrome and pneumonitis should be treated as having a correlation coefficient of 1 because it would not make sense for pneumonitis to require higher doses than the pulmonary syndrome.

7.0 FOOD-CHAIN PARAMETERS

A large number of food chain parameters are included in the COMIDA2 food-chain model. All of the parameters relevant to the COMIDA2 food-chain model that were evaluated in the expert elicitation are described in this section.

7.1 Distributions of Animal Feed Parameters

The first category of data evaluated in this section is the quantity of feed consumed by livestock that are most commonly consumed as meat or animal products in the United States. Three types of food consumption as well as consumption of soil by five categories of livestock are evaluated. All consumption rates are calculated on a dry basis. Cattle, sheep, and poultry are assumed to live predominantly outdoors; pigs are assumed to live predominantly in a barn. The values at 11 quantile levels and for the mean and mode are shown in Tables 7-1 and 7-2. Figures 7-1 through 7-15 show comparisons between the expert data and the resampled curves from which the data in Tables 7-1 and 7-2 are extracted.

It is not entirely obvious how to best interpolate the data in Tables 7-1 and 7-2 between quantile levels. Since some of the quantities in both of the tables are zero, it is probably best to use a simple linear interpolation.

Table 7-1: Daily consumption by beef and dairy cattle.

| Quantile | Dry Quantity Consumed (kg/dy) | | | | | | | |
| | Dairy Cattle | | | | Beef Cattle | | | |
	Pasture Grass	Silage/ Hay	Cereals	Soil	Pasture Grass	Silage/ Hay	Cereals	Soil
0.00	1.0	0.1	0.0	0.01	1.5	0.0	0.0	0.01
0.01	2.0	0.1	0.0	0.01	2.4	0.0	0.0	0.01
0.05	3.2	0.4	0.1	0.01	3.5	0.0	0.0	0.01
0.10	4.5	1.0	0.5	0.02	4.6	1.0	0.1	0.02
0.25	7.0	2.8	2.7	0.09	6.2	3.4	1.5	0.09
0.50	12.0	5.9	5.7	0.26	8.2	8.1	6.3	0.26
0.75	16.6	11.4	9.7	0.70	10.6	10.2	9.3	0.70
0.90	20.5	19.6	14.8	1.24	12.8	12.4	11.1	1.24
0.95	23.7	22.6	16.6	1.67	14.7	14.4	12.8	1.67
0.99	40.0	30.1	26.4	2.92	24.0	23.3	21.1	2.92
1.00	50.0	50.0	38.0	5.00	30.0	30.0	28.0	5.00
Mean	12.7	8.2	6.9	0.50	8.8	7.5	6.0	0.50
Mode	5.1	0.2	0.0	0.01	7.5	0.0	0.0	0.01

Table 7-2: Daily consumption by sheep, pigs, and poultry.

Quantile	Dry Quantity Consumed (kg/dy)							
	Sheep				Pigs		Poultry	
	Pasture Grass	Silage/ Hay	Cereals	Soil	Cereals	Soil	Cereals	Soil
0.00	0.20	0.00	0.00	0.01	0.3	0.00	0.015	0.000
0.01	0.27	0.00	0.00	0.01	0.5	0.00	0.026	0.000
0.05	0.45	0.00	0.00	0.01	0.9	0.00	0.044	0.000
0.10	0.56	0.15	0.10	0.02	1.1	0.00	0.054	0.000
0.25	0.86	0.44	0.44	0.05	1.5	0.01	0.074	0.001
0.50	1.29	0.97	0.86	0.12	2.3	0.03	0.101	0.005
0.75	1.81	1.66	1.44	0.21	10.8	0.14	0.134	0.015
0.90	2.29	2.29	1.94	0.33	13.7	0.32	0.172	0.034
0.95	2.73	2.55	2.36	0.45	14.7	0.41	0.196	0.049
0.99	4.52	4.46	3.89	0.75	20.4	0.75	0.303	0.096
1.00	6.00	6.00	5.00	1.00	32.0	1.00	0.404	0.160
Mean	1.42	1.14	1.01	0.16	5.2	0.10	0.11	0.013
Mode	1.06	0.00	0.00	0.03	1.3	0.00	0.10	0.000

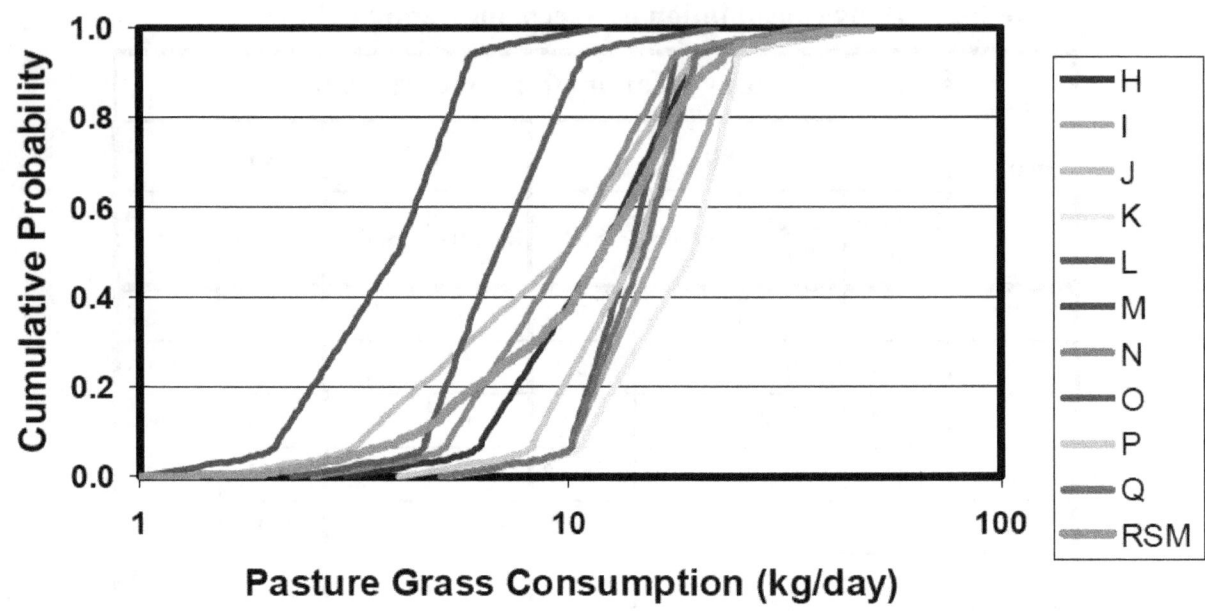

Figure 7-1.   Expert and resampled data for consumption rate (kg/day on a dry basis) of pasture grass by dairy cattle

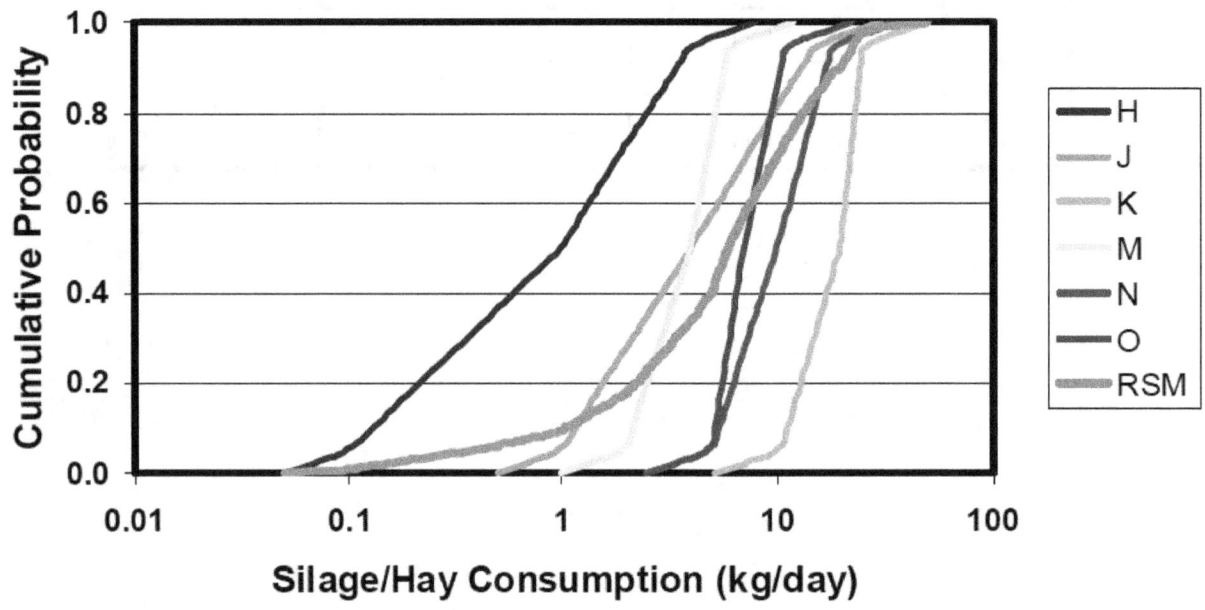

Figure 7-2.   Expert and resampled data for consumption rate (kg/day on a dry basis) of silage/hay by dairy cattle

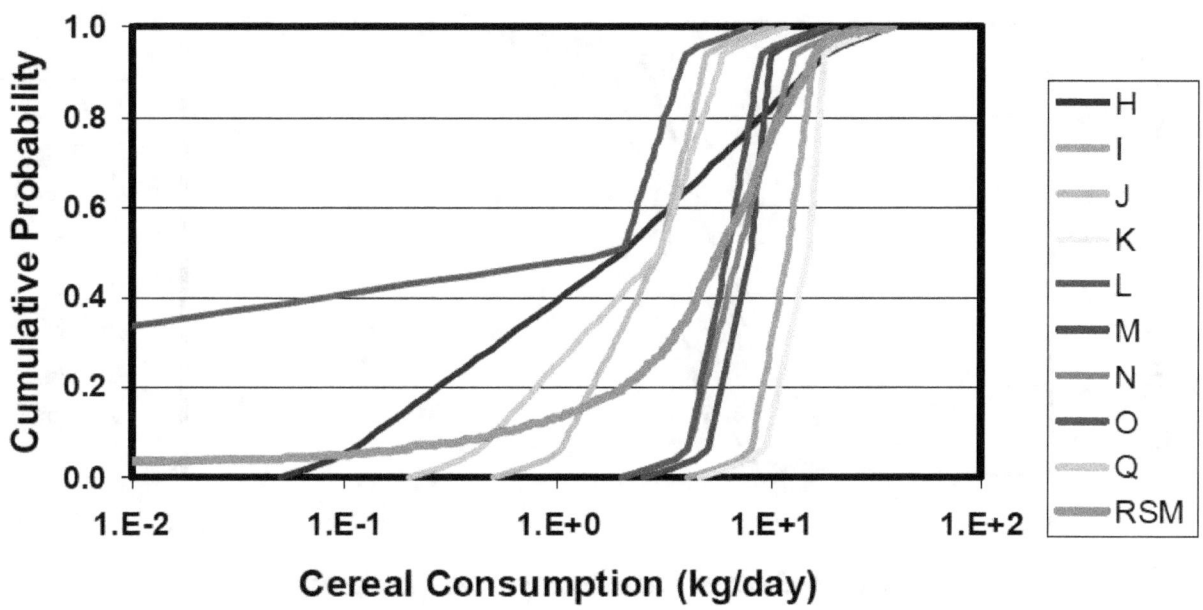

Figure 7-3. Expert and resampled data for consumption rate (kg/day on a dry basis) of cereals by dairy cattle

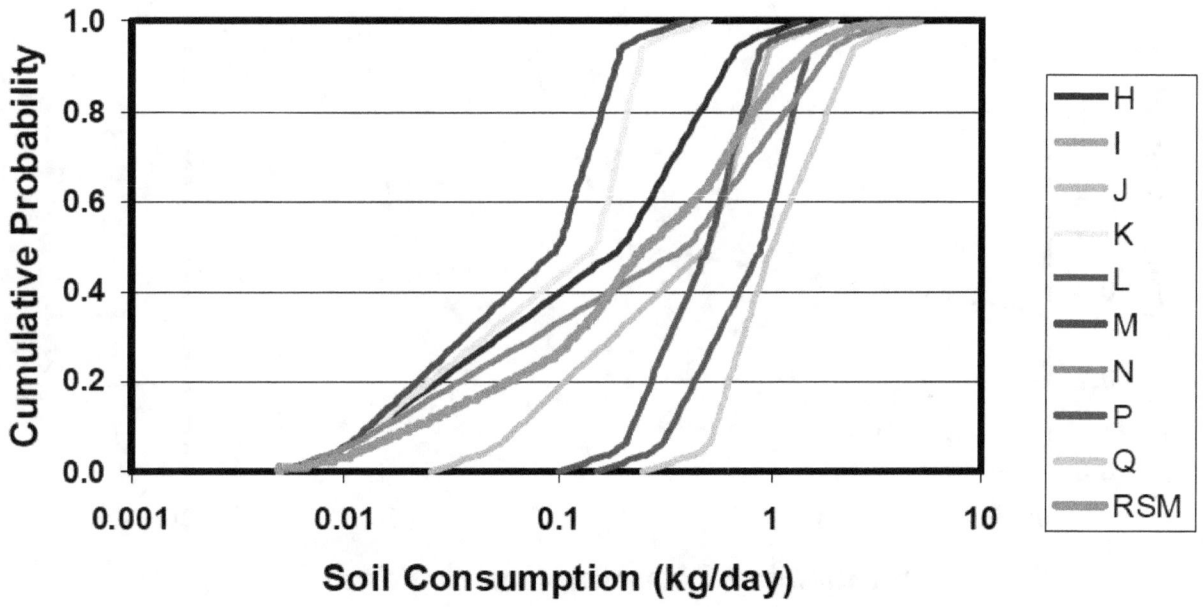

Figure 7-4. Expert and resampled data for consumption rate (kg/day on a dry basis) of soil by dairy and beef cattle

103

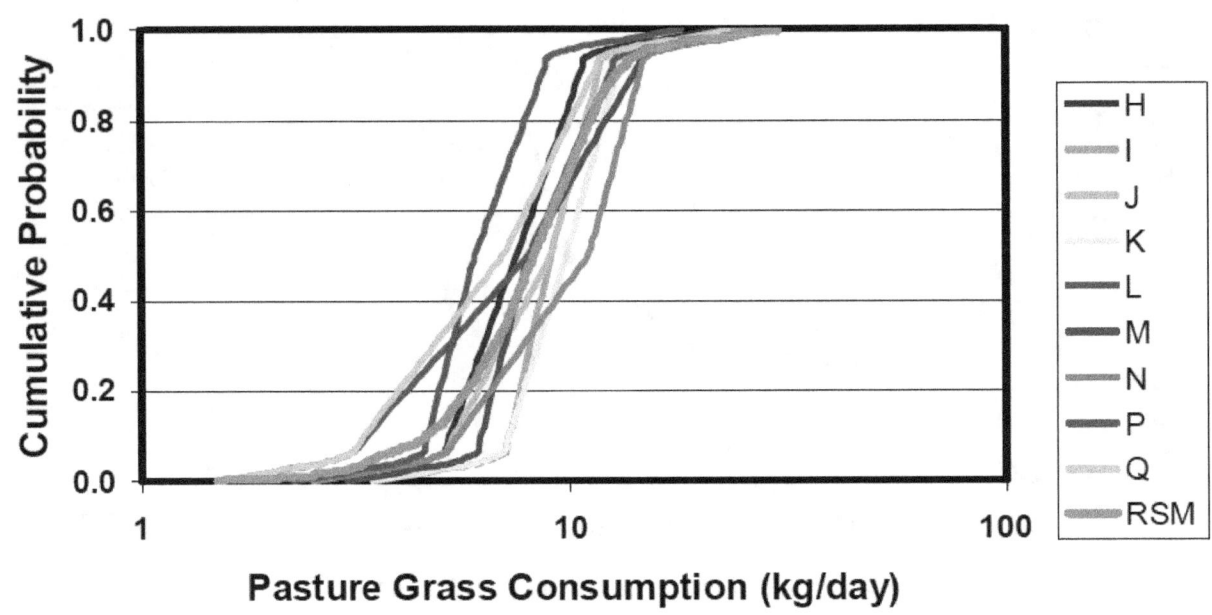

Figure 7-5.   Expert and resampled data for consumption rate (kg/day on a dry basis) of pasture grass by beef cattle

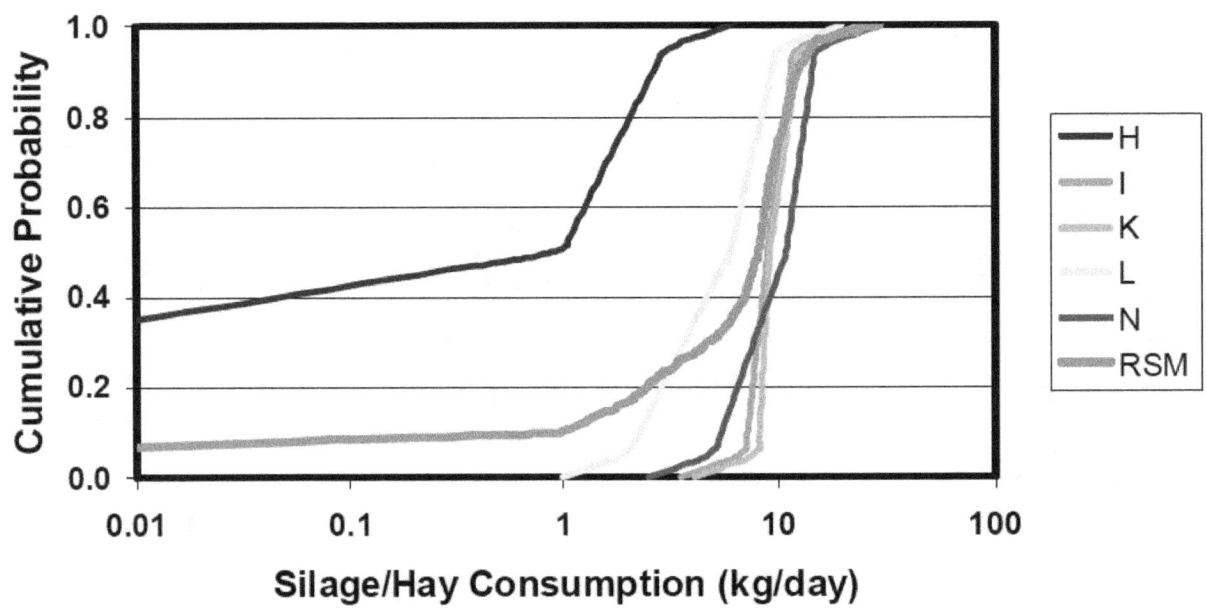

Figure 7-6.   Expert and resampled data for consumption rate (kg/day on a dry basis) of silage/hay by beef cattle

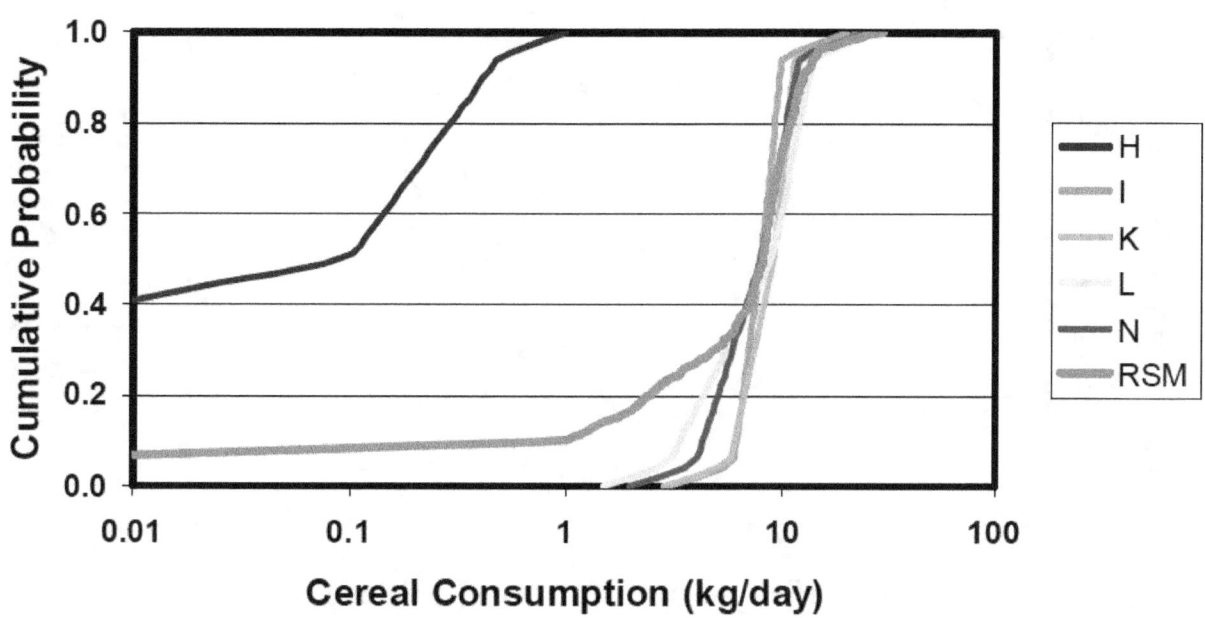

Figure 7-7. Expert and resampled data for consumption rate (kg/day on a dry basis) of cereals by beef cattle

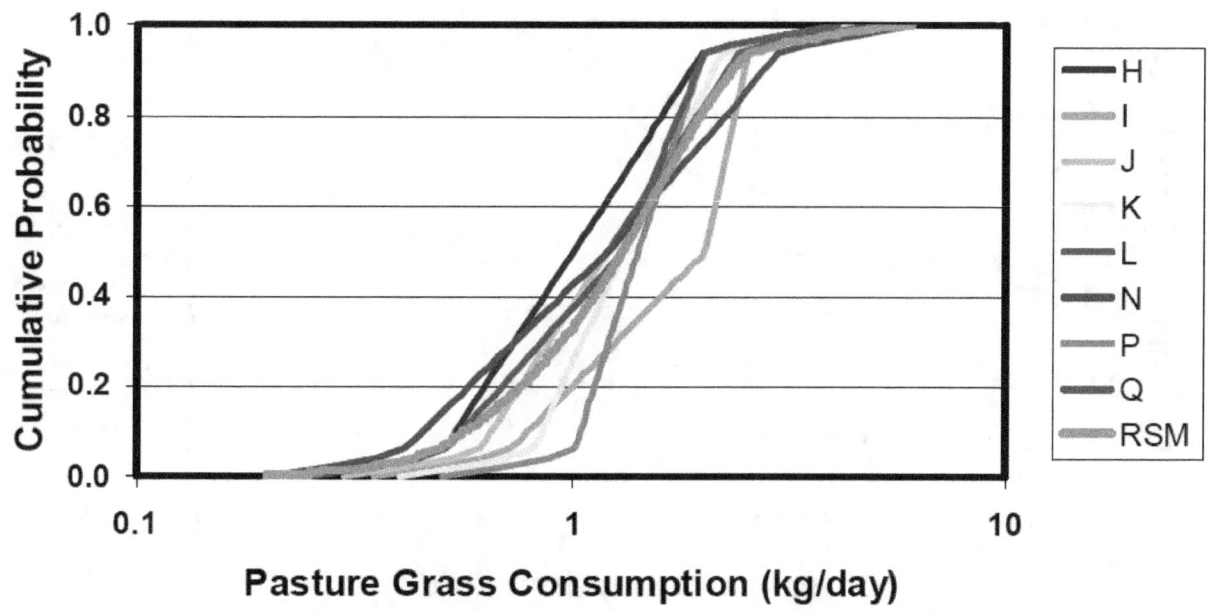

Figure 7-8. Expert and resampled data for consumption rate (kg/day on a dry basis) of pasture grass by sheep

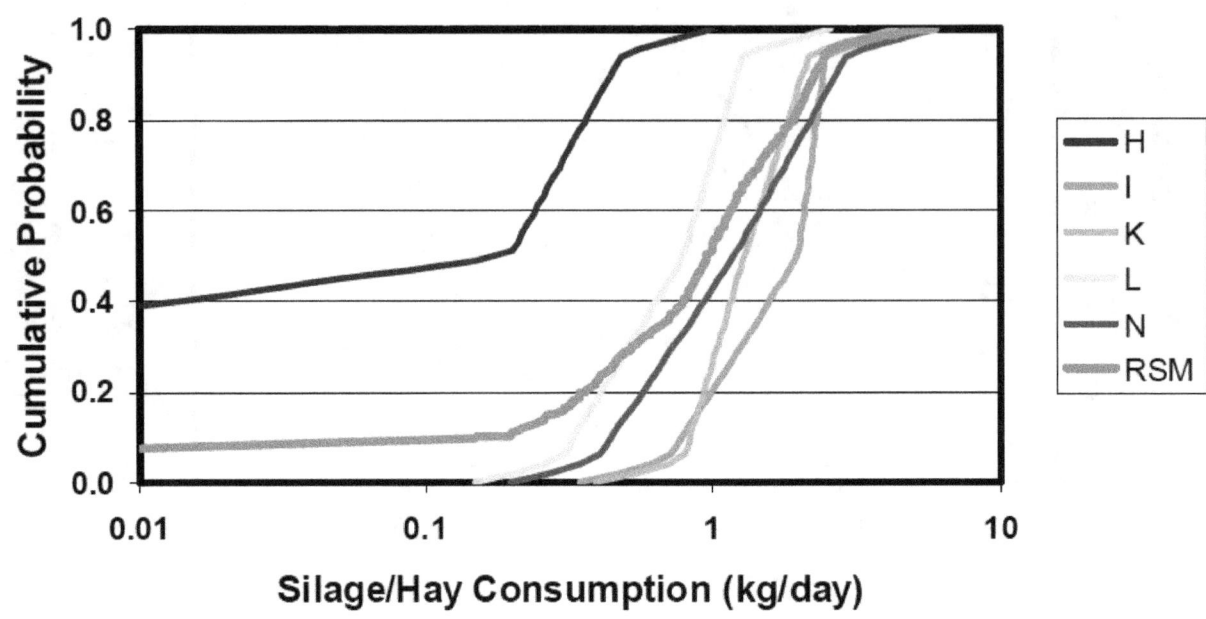

Figure 7-9.   Expert and resampled data for consumption rate (kg/day on a dry basis) of silage/hay by sheep

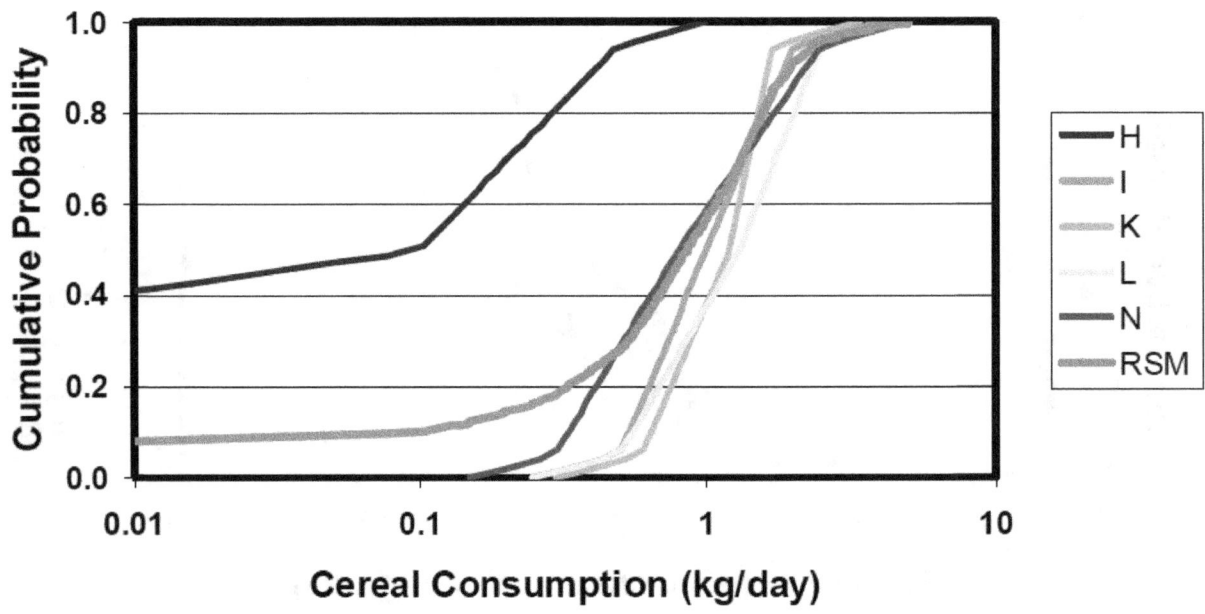

Figure 7-10.   Expert and resampled data for consumption rate (kg/day on a dry basis) of cereals by sheep

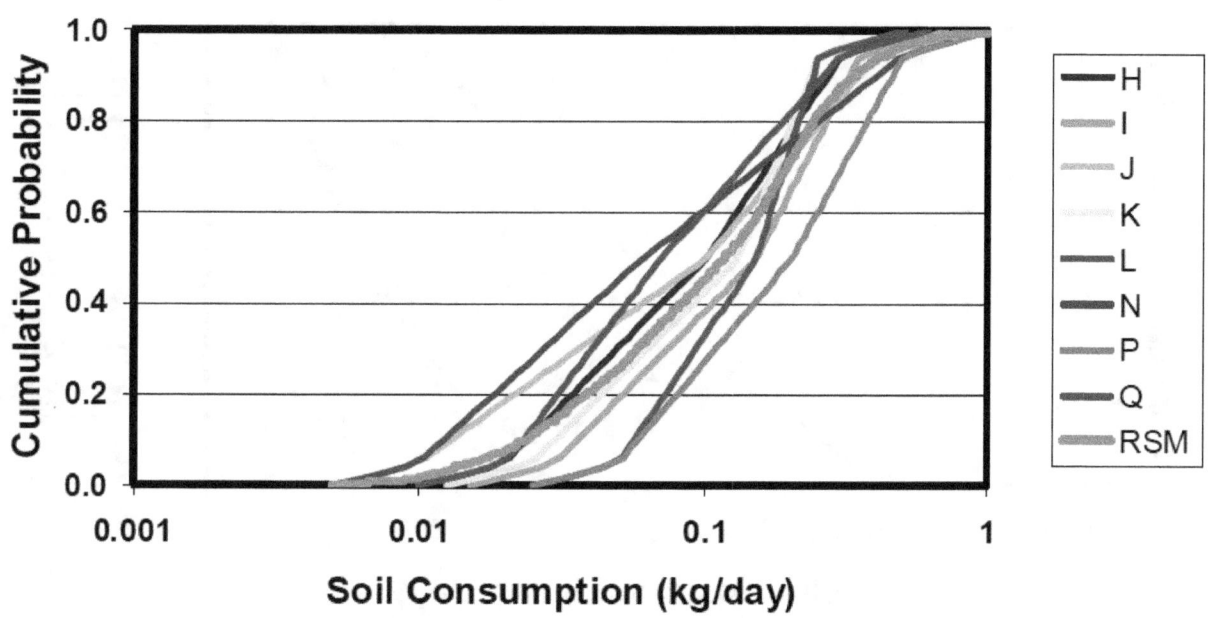

Figure 7-11. Expert and resampled data for consumption rate (kg/day on a dry basis) of soil by sheep

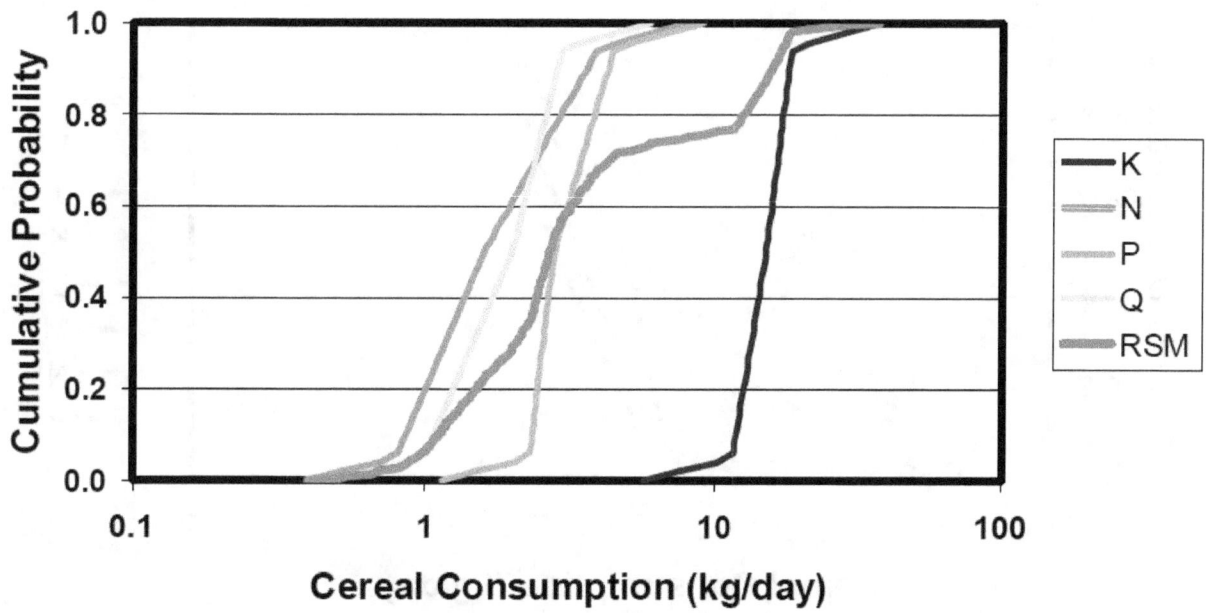

Figure 7-12. Expert and resampled data for consumption rate (kg/day on a dry basis) of cereal by pigs

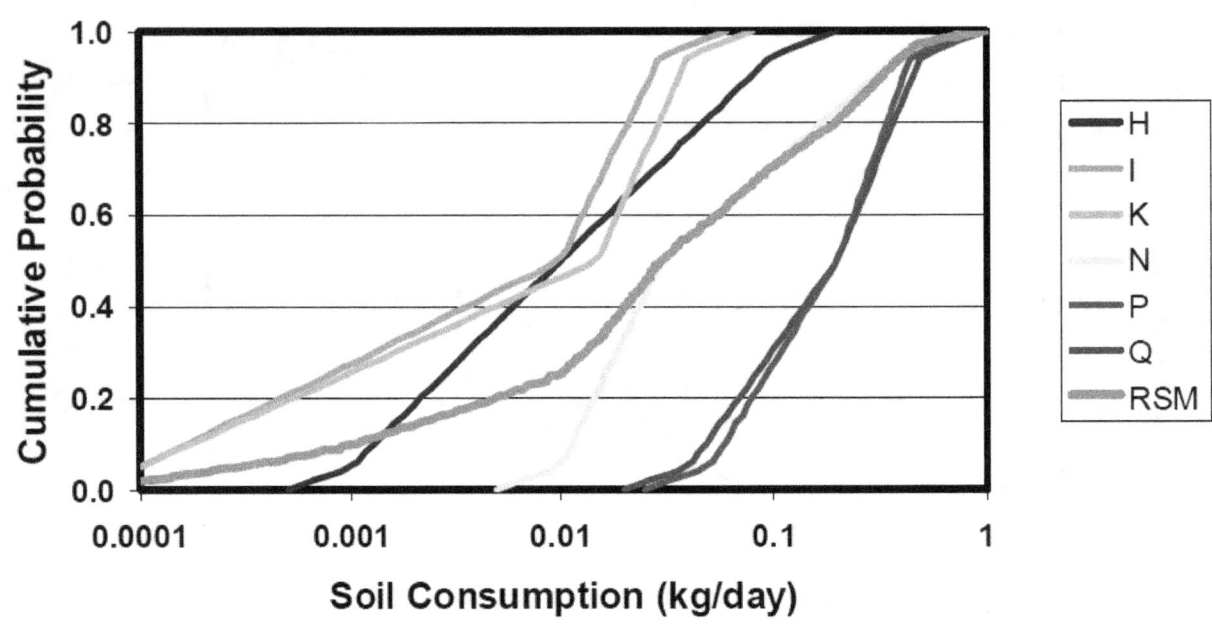

Figure 7-13. Expert and resampled data for consumption rate (kg/day on a dry basis) of soil by pigs

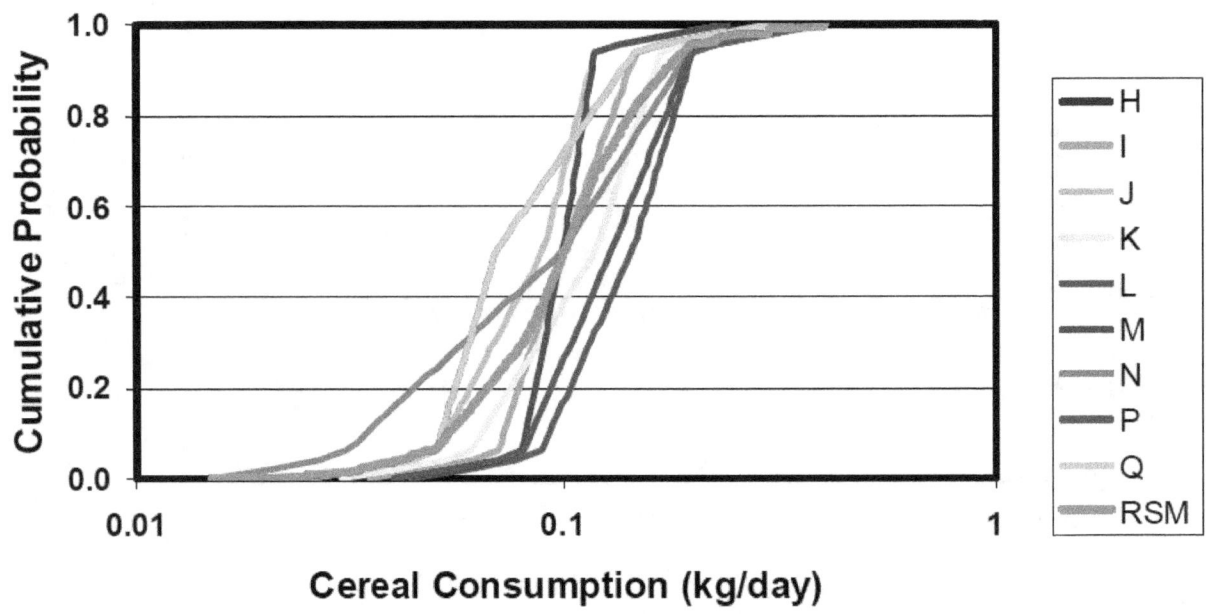

Figure 7-14. Expert and resampled data for consumption rate (kg/day on a dry basis) of cereal by poultry

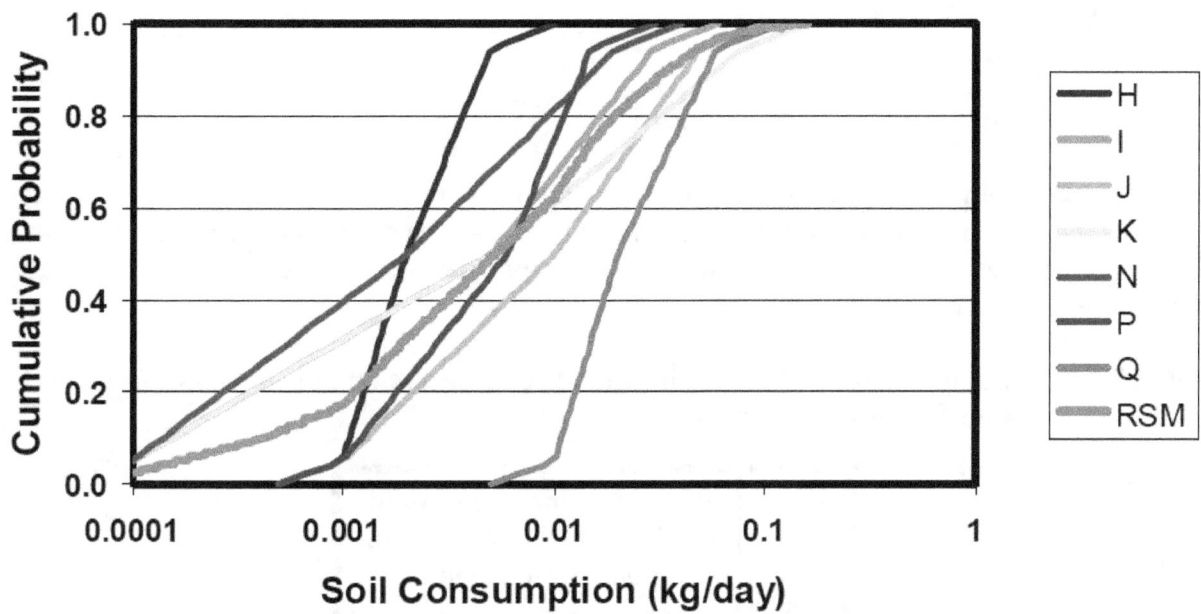

Figure 7-15. Expert and resampled data for consumption rate (kg/day on a dry basis) of soil by poultry

7.2 Correlations Between Animal Feed Parameters

It is clear that the quantities in Tables 7-1 and 7-2 for a specified livestock type must be correlated. For example, a dairy cow that consumes 50 kg/dy of pasture grass would not also consume the maximum quantities of hay and cereals. So, the various categories of food consumption must be negatively correlated. The correlations are made more difficult by the fact that there are three food categories from which the total should be more or less constant. Also, consumption of soil by cattle and sheep is likely to be correlated with consumption of pasture grass and to a lesser extent silage/hay; cereals are likely to be eaten from some type of feed trough that would limit the consumption of soil. As a suggestion, the correlation between pasture grass and hay consumption should be strongly inversely correlated (i.e., -0.9 to -1.0) for both dairy and beef cattle. The correlation between grains and the other two feed categories should be weak (i.e., -0.5 to 0.5). The correlation between soil and the three feed categories should also be weak.

Consumption of cereals by pigs and poultry is even more difficult to interpret. The expert elicitation process did not prescribe an alternative food source to grains, but the animals are presumably given some other type of feed if grain is not available. The only other type of feed allowed for poultry in COMIDA2 is legumes, but this would not normally be a significant portion of the food source for chickens. Thus, the large uncertainty in the daily intake of grains by chickens (and also pigs) seems difficult to justify. Since no alternative food source is suggested, it is not obvious whether or not the daily intake of soil by chickens (and pigs) should be correlated with the intake of grains. Therefore, a weak correlation (i.e., -0.5 to 0.5) between consumption of grains and soil is reasonable.

7.3   Distributions of Concentration Ratios

The second category of data evaluated in this section is the ratio of grain and root concentrations to elemental concentrations in the surrounding soil. The values at 11 quantile levels are shown in Table 7-3. Figures 7-16 through 7-19 show comparisons between the expert data and the resampled curves from which the data in Table 7-3 are extracted.

Table 7-3:   Ratio of concentration in foodstuff to concentration in soil (Bq/kg fresh plant mass / Bq/kg dry soil mass).

| Quantile | Ratio of Concentration in Foodstuff to Concentration in Soil (Dimensionless) | | | |
| | Grain | | Root | |
	Sr	Cs	Sr	Cs
0.00	0.00	0.00	0.000	0.00000
0.01	0.00	0.00	0.000	0.00000
0.05	0.01	0.01	0.000	0.00000
0.10	0.01	0.02	0.000	0.00000
0.25	0.04	0.03	0.000	0.00000
0.50	0.13	0.07	0.006	0.00000
0.75	0.29	0.16	0.017	0.00007
0.90	0.42	0.33	0.075	0.00086
0.95	0.93	0.50	0.121	0.00197
0.99	2.73	0.91	0.197	0.00454
1.00	5.25	1.73	0.347	0.00600
Mean	0.27	0.14	0.023	0.00032
Mode	0.01	0.03	0.000	0.00000

110

It is not entirely obvious how to best interpolate the data in Table 7-3 between quantile levels. Since some of the quantities are zero, it is probably best to use a simple linear interpolation.

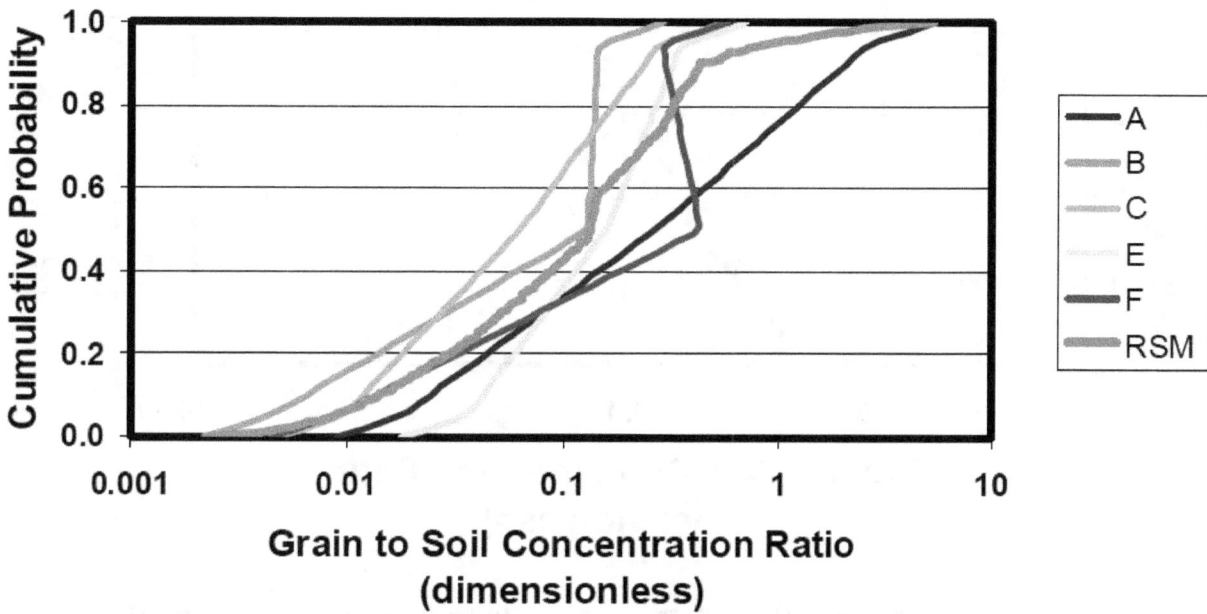

Figure 7-16. **Expert and resampled data for grain to soil concentration ratio (Bq/kg Fresh Plant Mass / Bq/kg Dry Soil Mass) for Sr**

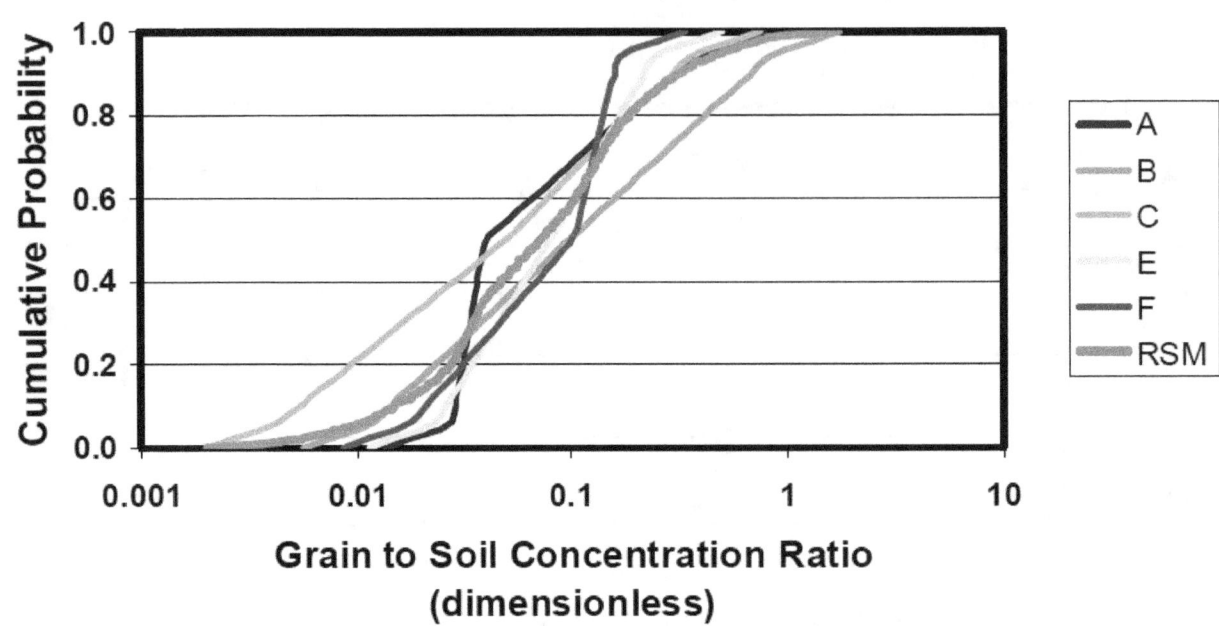

Figure 7-17.   Expert and resampled data for grain to soil concentration ratio (Bq/kg Fresh Plant Mass / Bq/kg Dry Soil Mass) for Cs

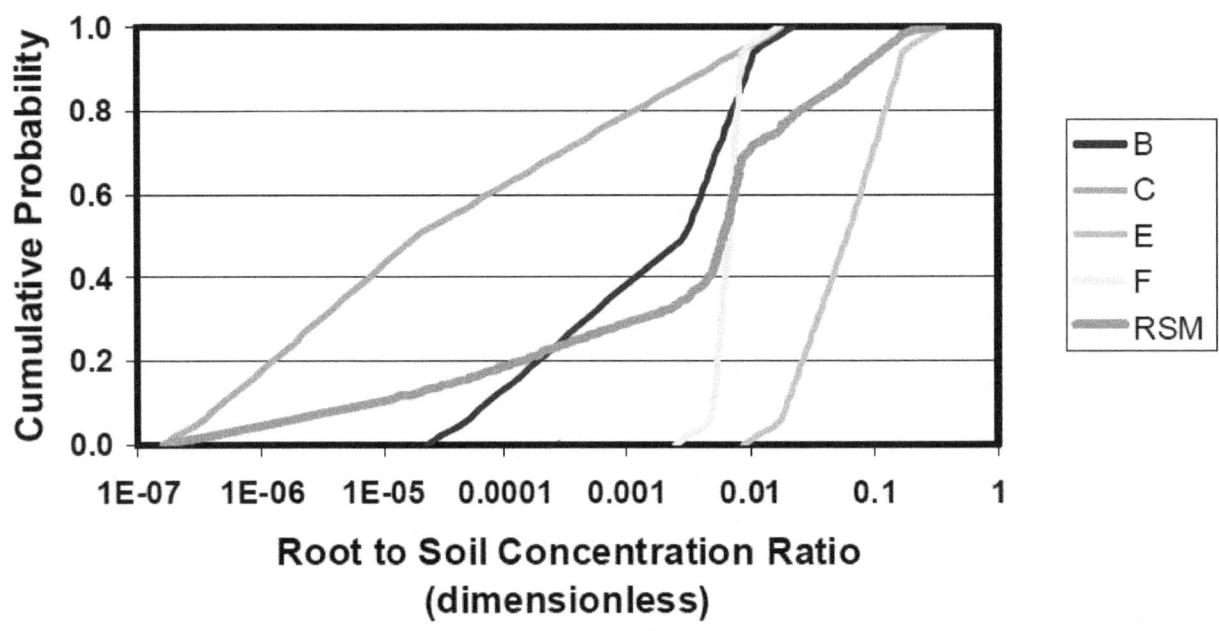

Figure 7-18.   Expert and resampled data for root to soil concentration ratio (Bq/kg Fresh Plant Mass / Bq/kg Dry Soil Mass) for Sr

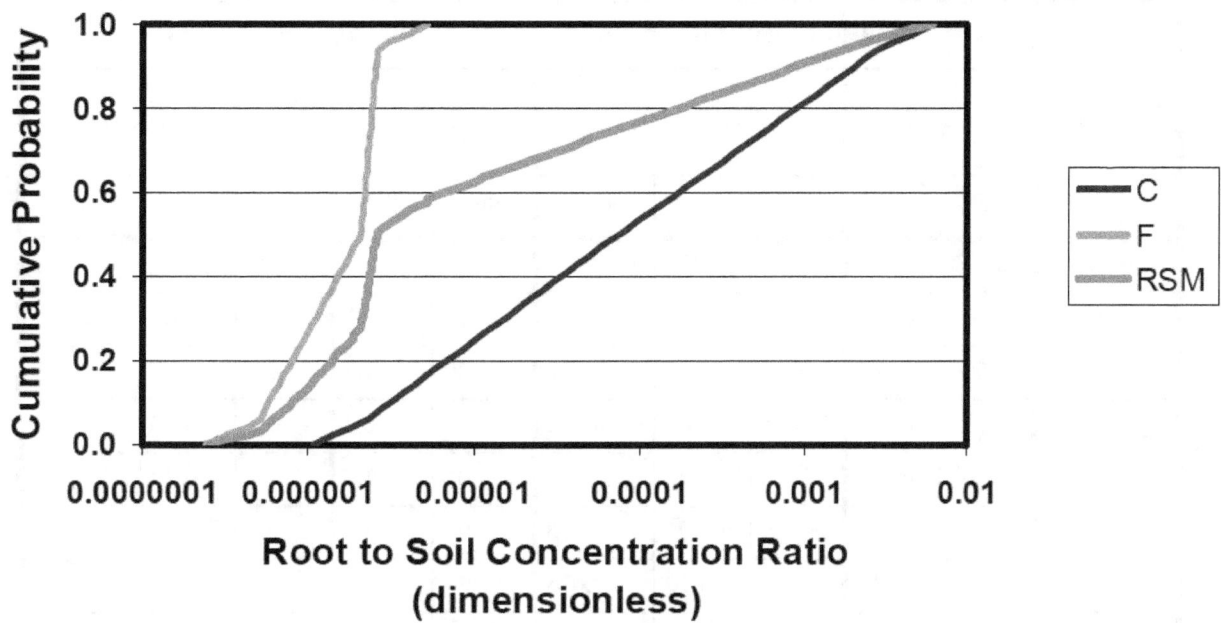

Figure 7-19. Expert and resampled data for root to soil concentration ratio (Bq/kg Fresh Plant Mass / Bq/kg Dry Soil Mass) for Cs

7.4   Correlations Between Concentration Ratios

Since the chemical properties of cesium and strontium are different, there is no reason to assume that the concentration ratios for these elements should be correlated. There could be some correlation between the uptake of these elements by grains and roots, but the only defensible assumption is that they are also uncorrelated. Thus, it is reasonable that these four distributions be treated as completely uncorrelated.

7.5   Distributions of Transfer Coefficients

The third category of data evaluated in this section is a set of transfer coefficients. These are quantities with dimensions of d/kg or d/l that reflect the equilibrium ratio of the activity in foodstuffs to the daily activity intake of the animal generating the foodstuff. Foodstuffs considered are beef, milk, poultry, and eggs. The elements considered are strontium and cesium; iodine is also included for milk and eggs. The values at 11 quantile levels are shown in Table 7-4. Figures 7-20 through 7-29 show comparisons between the expert data and the resampled curves from which the data in Table 7-4 are extracted.

Table 7-4:   Ratio of concentration in foodstuff to daily ingestion level.

Quantile	Ratio of Activity in Foodstuff to Daily Ingestion Level									
	Beef (d/kg)		Milk (d/l)			Poultry (d/kg)		Eggs (d/kg)		
	Sr	Cs	Sr	Cs	I	Sr	Cs	Sr	Cs	I
0.00	0.000	0.001	0.0002	0.000	0.000	0.00	0.2	0.0	0.03	0.0
0.01	0.000	0.003	0.0002	0.000	0.001	0.00	0.2	0.0	0.03	0.0
0.05	0.000	0.007	0.0004	0.001	0.001	0.00	0.4	0.0	0.07	0.0
0.10	0.000	0.009	0.0005	0.001	0.001	0.01	0.7	0.0	0.10	0.3
0.25	0.001	0.018	0.0010	0.002	0.003	0.02	1.6	0.1	0.22	1.4
0.50	0.003	0.039	0.0019	0.005	0.006	0.06	3.9	0.2	0.48	2.6
0.75	0.008	0.057	0.0032	0.010	0.015	0.30	8.9	0.4	0.80	3.8
0.90	0.011	0.088	0.0042	0.020	0.027	1.05	12.0	0.9	1.59	5.4
0.95	0.016	0.120	0.0056	0.030	0.036	2.29	15.9	2.5	2.03	7.7
0.99	0.079	0.196	0.0100	0.138	0.068	6.03	25.6	9.8	3.31	11.4
1.00	0.200	0.400	0.0140	0.400	0.100	8.00	50.0	17.2	5.00	20.0
Mean	0.006	0.046	0.0023	0.011	0.011	0.44	5.9	0.6	0.67	3.0
Medium	0.000	0.053	0.0007	0.001	0.001	0.01	0.5	0.0	0.07	0.0

It is not entirely obvious how to best interpolate the data in Table 7-4 between quantile levels. Since some of the quantities are zero, it is probably best to use a simple linear interpolation.

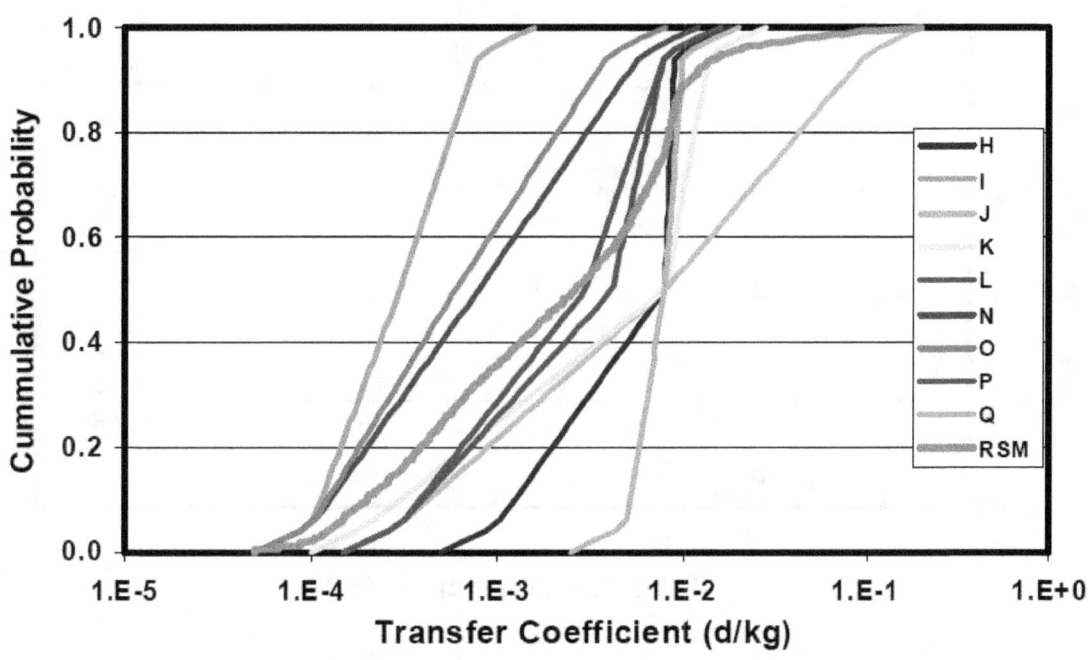

Figure 7-20. Expert and resampled data for beef transfer coefficient at steady state (Bq/kg / Bq/d) for strontium

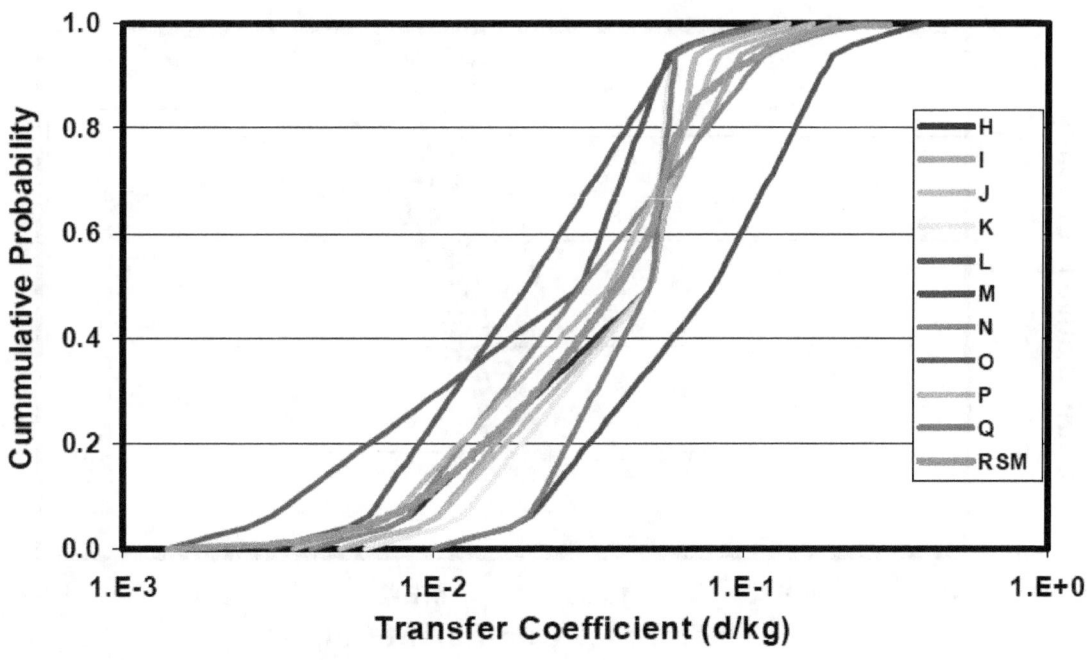

Figure 7-21. Expert and resampled data for beef transfer coefficient at steady state (Bq/kg / Bq/d) for cesium

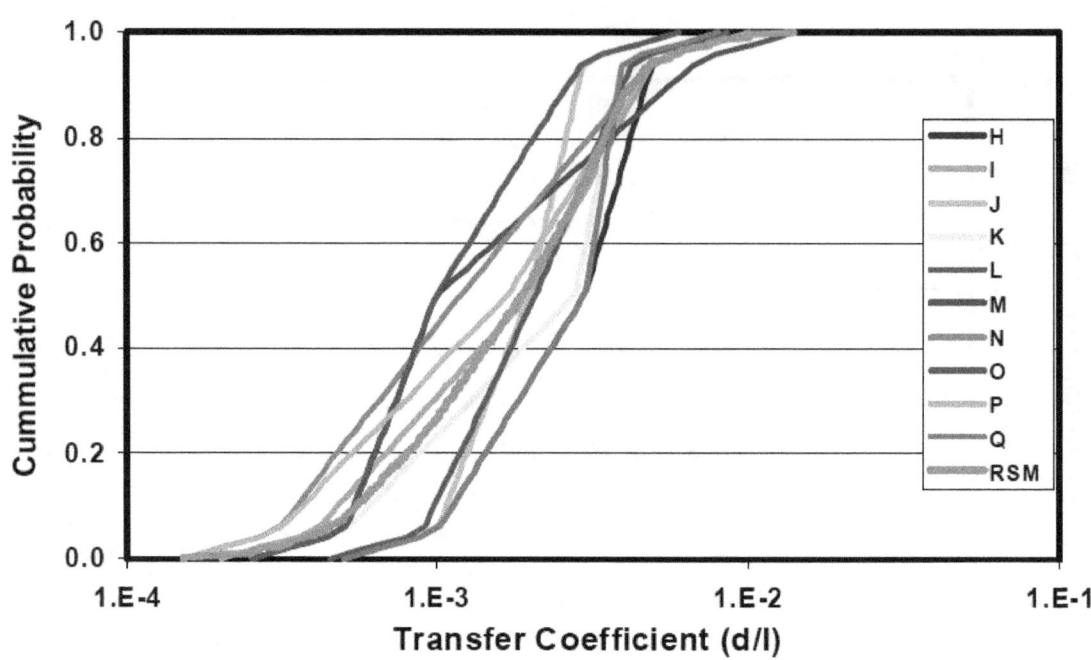

Figure 7-22. Expert and resampled data for milk transfer coefficient at steady state (Bq/l / Bq/d) for strontium

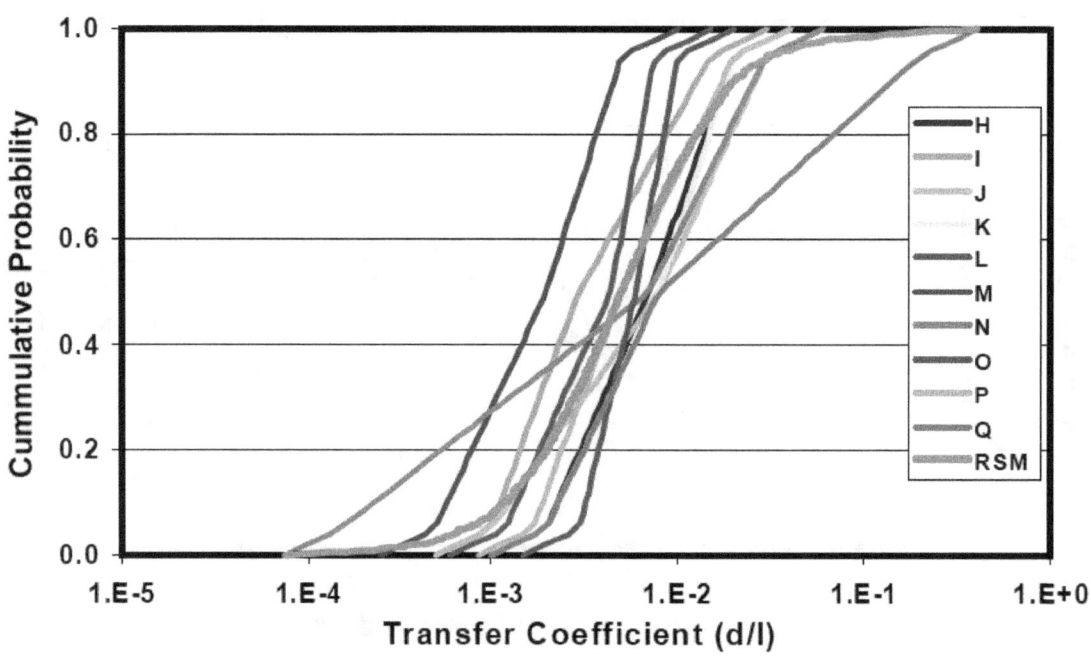

Figure 7-23. Expert and resampled data for milk transfer coefficient at steady state (Bq/l / Bq/d) for cesium

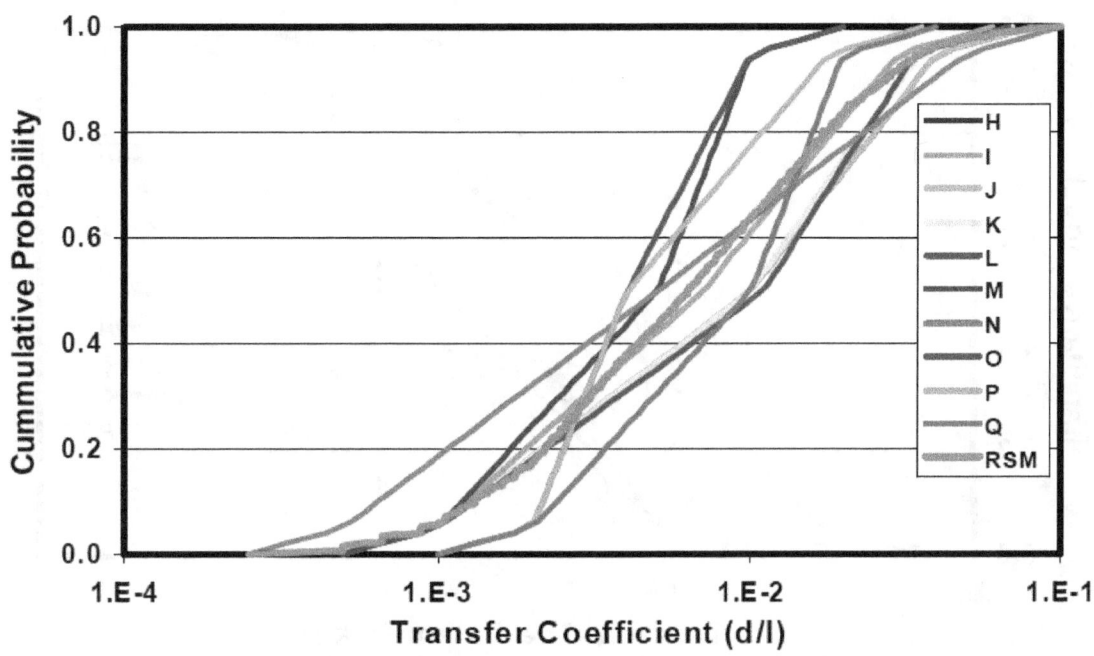

Figure 7-24. Expert and resampled data for milk transfer coefficient at steady state (Bq/l / Bq/d) for iodine

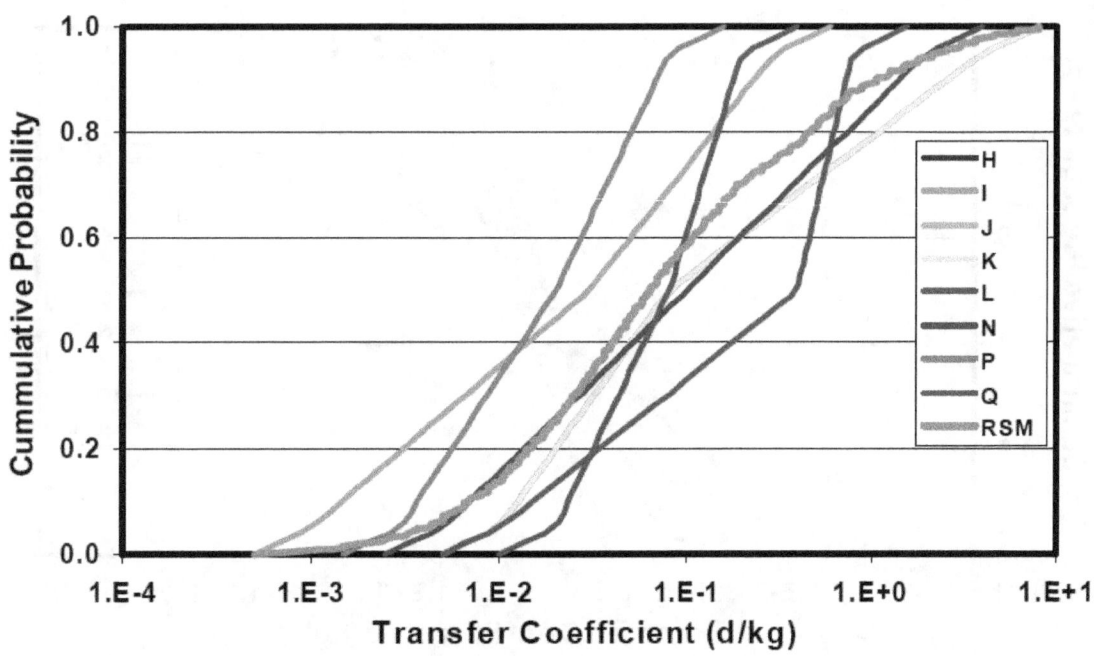

Figure 7-25. Expert and resampled data for poultry transfer coefficient at steady state (Bq/ kg / Bq/d) for strontium

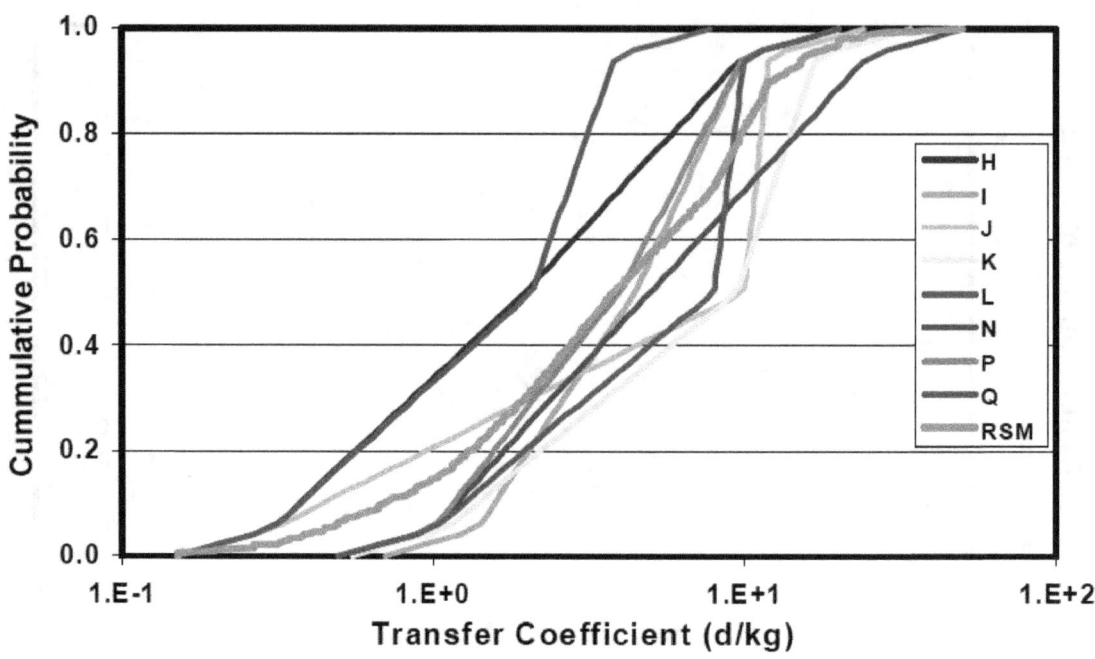

Figure 7-26. Expert and resampled data for poultry transfer coefficient at steady state (Bq/kg / Bq/d) for cesium

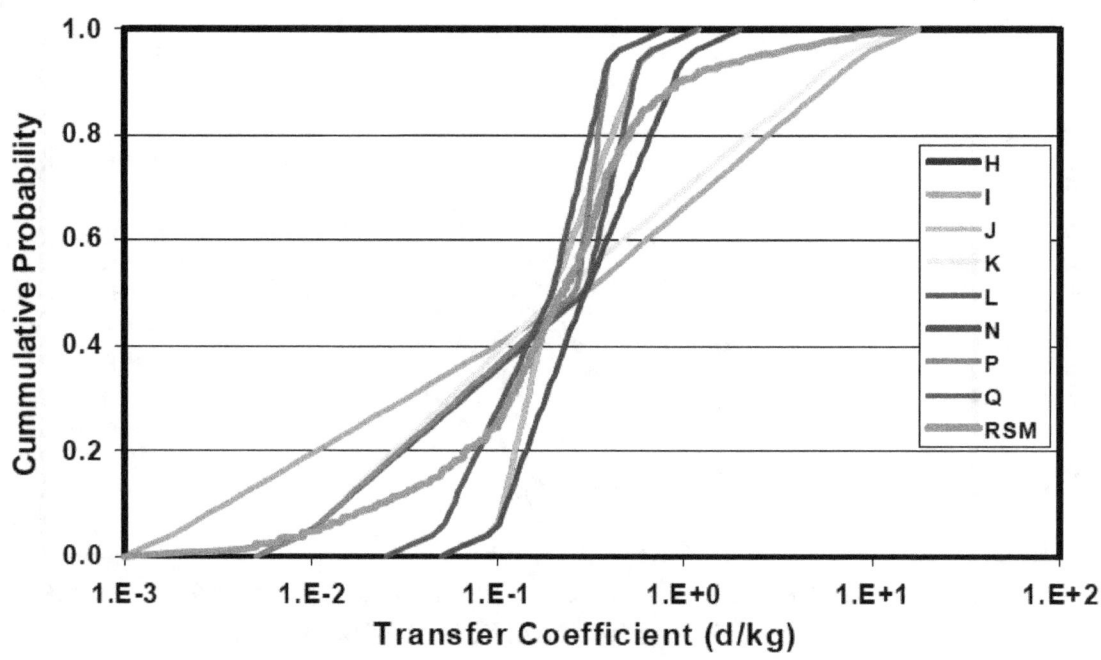

Figure 7-27. Expert and resampled data for egg transfer coefficient at steady state (Bq/kg / Bq/d) for strontium

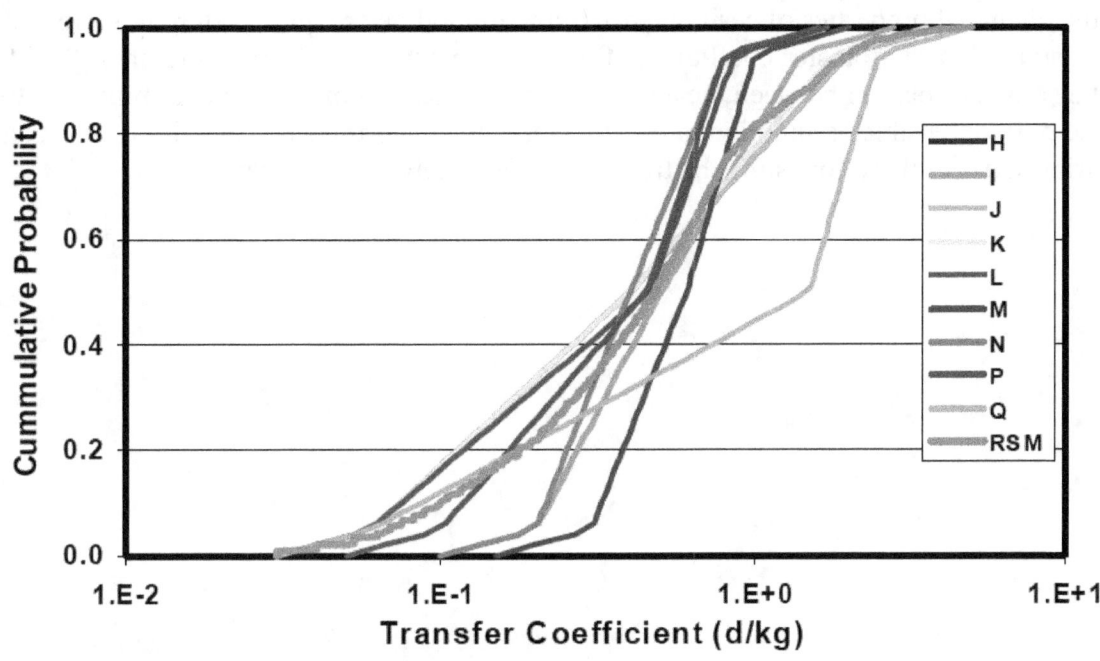

Figure 7-28. Expert and resampled data for egg transfer coefficient at steady state (Bq/kg / Bq/d) for cesium

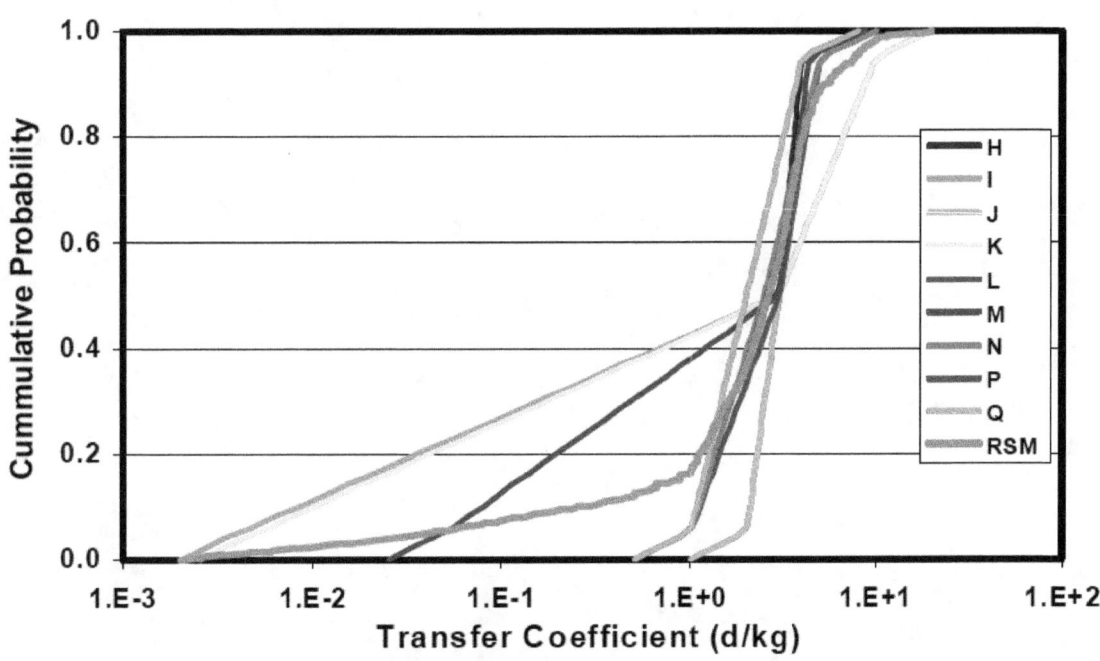

Figure 7-29. Expert and resampled data for egg transfer coefficient at steady state (Bq/kg / Bq/d) for iodine

7.6   Correlations Between Transfer Coefficients

Since the chemical properties of cesium, strontium, and iodine are quite different, there is no reason to assume that the transfer coefficients for these elements should be correlated. While there could be some correlation between the accumulation of these elements in beef, milk, poultry, and eggs, the only defensible assumption is that they are also uncorrelated. Thus, it is reasonable that the distributions for these foodstuffs be treated as completely uncorrelated.

8.0 CONCLUSIONS

The results of the expert elicitation conducted by the NRC and the CEC have been evaluated to provide ranges of values and degrees of belief for parameters in off-site consequence analyses that are likely to have significant or moderate influence on the calculated results. This present work represents the effort to fairly represent the divergent opinions of the experts while maintaining the resulting parameters within physical limits. This evaluation also includes correlation coefficients between variables when appropriate. The medians, means, and modes of the distributions are presented so that the user can choose a value when only a point estimate calculation is needed.

9.0 REFERENCES

1. F.T. Harper, et al., "Probabilistic Accident Consequence Uncertainty Analysis, Dispersion and Deposition Uncertainty Assessment," NUREG/CR-6244, NRC, Washington, DC, January 1995.

2. F.T. Harper, et al., "Probabilistic Accident Consequence Uncertainty Analysis, Food Chain Uncertainty Assessment," NUREG/CR-6523, NRC, Washington, DC, June 1997.

3. F.T. Harper, et al., "Probabilistic Accident Consequence Uncertainty Analysis, Uncertainty Assessment for Deposited Material and External Doses," NUREG/CR-6526, NRC, Washington, DC, December 1997.

4. F.T. Harper, et al., "Probabilistic Accident Consequence Uncertainty Analysis, Early Health Effects Uncertainty Assessment," NUREG/CR-6545, NRC, Washington, DC, December 1997.

5. F.T. Harper, et al., "Probabilistic Accident Consequence Uncertainty Analysis, Late Health Effects Uncertainty Assessment," NUREG/CR-6555, NRC, Washington, DC, December 1997.

6. F.T. Harper, et al., "Probabilistic Accident Consequence Uncertainty Analysis, Uncertainty Assessment for Internal Dosimetry," NUREG/CR-6571, NRC, Washington, DC, April 1998.

7. D. Chanin and M.L. Young, "Code Manual for MACCS2, User's Guide," NUREG/CR-6613, NRC, Washington, DC, May 1998.

8. T. A. Wheeler, G. D. Wyss, and F. T. Harper, "Cassini Spacecraft Uncertainty Analysis Data and Methodology Review and Update, Volume 1: Updated Parameter Uncertainty Models for Consequence Analysis," SAND2000-2719/1, Sandia National Laboratories, Albuquerque, NM, 2000.

9. D. B. Turner, "Workbook of Atmospheric Dispersion Estimates, An Introduction to Dispersion Modeling," 2nd Ed., Lewis Publishers, CRC Press, Boca Raton, 1994.

U.S. NUCLEAR REGULATORY COMMISSION

BIBLIOGRAPHIC DATA SHEET

(See instructions on the reverse)

1. REPORT NUMBER (Assigned by NRC, Add Vol., Supp., Rev., and Addendum Numbers, if any.)
NUREG/CR-7161

2. TITLE AND SUBTITLE

Synthesis of Distributions Representing Important Non-Site-Specific Parameters in Offsite Consequence Analyses

3. DATE REPORT PUBLISHED

MONTH	YEAR
April	2013

4. FIN OR GRANT NUMBER

5. AUTHOR(S)

N.E. Bixler, E. Clauss, and C.W. Morrow

6. TYPE OF REPORT

NUREG/CR

7. PERIOD COVERED (Inclusive Dates)

8. PERFORMING ORGANIZATION - NAME AND ADDRESS (If NRC, provide Division, Office or Region, U. S. Nuclear Regulatory Commission, and mailing address; if contractor, provide name and mailing address.)

Sandia National Laboratories
Albuquerque, New Mexico 87185-0734
Operated for the U.S. Department of Energy

9. SPONSORING ORGANIZATION - NAME AND ADDRESS (If NRC, type "Same as above", if contractor, provide NRC Division, Office or Region, U. S. Nuclear Regulatory Commission, and mailing address.)

Division of Systems Analysis, Office of Nuclear Regulatory Research
U.S. Nuclear Regulatory Commission
Washington, DC 20555-0001

10. SUPPLEMENTARY NOTES

11. ABSTRACT (200 words or less)

The United States and the Commission of European Communities conducted a series of expert elicitations to obtain distributions for uncertain variables used in health consequence analyses related to accidental release of nuclear material. The distributions reflect degrees of belief for non-site-specific parameters that are uncertain and are likely to have significant or moderate influence on the results. The present work presents the effort to develop ranges of values and degrees of belief that fairly represent the divergent opinions of the experts while maintaining the resulting parameters within physical limits. Where necessary, there is a discussion of correlation coefficients that should be included when the uncertainty is used in a calculation. The methodology used a resampling of the experts' values and was based on the assumption of equal weights of the experts' opinions. Various statistical properties of the distributions, the median, the mean, and the mode, are presented so that the user can choose a parameter value when only a point estimate is desired.

12. KEY WORDS/DESCRIPTORS (List words or phrases that will assist researchers in locating the report.)

MACCS2
WinMACCS
Uncertain Parameters
CEC
Uncertainty Analysis
Offsite Consequence Analysis

13. AVAILABILITY STATEMENT

unlimited

14. SECURITY CLASSIFICATION

(This Page)

unclassified

(This Report)

unclassified

15. NUMBER OF PAGES

16. PRICE

NUREG/CR-7161

Synthesis of Distributions Representing Important Non-Site-Specific Parameters in Off-Site Consequence Analyses

April 2013

www.ingramcontent.com/pod-product-compliance
Lightning Source LLC
Chambersburg PA
CBHW080252180526
45167CB00006B/2499